Heinrich Behrens, John W. Judd

A Manual of microchemical Analysis

Heinrich Behrens, John W. Judd

A Manual of microchemical Analysis

ISBN/EAN: 9783337139971

Printed in Europe, USA, Canada, Australia, Japan

Cover: Foto ©berggeist007 / pixelio.de

More available books at **www.hansebooks.com**

A MANUAL

OF

MICROCHEMICAL ANALYSIS

BY

PROFESSOR H. BEHRENS
OF THE POLYTECHNIC SCHOOL IN DELFT, HOLLAND

WITH AN INTRODUCTORY CHAPTER

BY

PROFESSOR JOHN W. JUDD, F.R.S.
OF THE ROYAL COLLEGE OF SCIENCE, LONDON

WITH 84 ILLUSTRATIONS DRAWN BY THE AUTHOR

London
MACMILLAN AND CO.
AND NEW YORK
1894

CONTENTS

	PAGE
INTRODUCTION	xiii

PART I

GENERAL METHOD AND REACTIONS

I. HISTORICAL REMARKS	1
II. AIM OF MICROCHEMICAL ANALYSIS	4
III. APPARATUS	12
IV. REAGENTS	21
V. REACTIONS	29
§ 1. Potassium	29
§ 2. Sodium	31
§ 3. Lithium	34
(Ammonium, see Nitrogen, § 53, *b*)	35
§ 4. Cæsium	35
§ 5. Rubidium	37
§ 6. Thallium	38
§ 7. Silver	40
§ 8. Magnesium	42
§ 9. Beryllium	44
§ 10. Manganese	46
§ 11. Cobalt	48
§ 12. Nickel	51
§ 13. Zinc	52
§ 14. Cadmium	56
§ 15. Cerium	58

	PAGE
§ 16. Lanthanum	59
§ 17. Didymium	60
§ 18. Yttrium and Erbium	62
§ 19. Barium	63
§ 20. Strontium	66
§ 21. Calcium	70
§ 22. Lead	74
§ 23. Copper	78
§ 24 and § 25. Mercury—	
(1) *Mercurous Compounds*	81
(2) *Mercuric Compounds*	82
§ 26. Gold	83
§ 27 and § 28. Platinum—	
(1) *Platinum Bichloride (Platinous Chloride)*	85
(2) *Platinum Tetrachloride (Chloroplatinic Acid)*	86
§ 29. Palladium	87
§ 30. Iridium	89
§ 31. Rhodium	89
§ 32. Ruthenium	90
§ 33. Osmium	91
§ 34 and § 35. Tin—	
(1) *Stannous Chloride*	92
(2) *Stannic Chloride*	93
§ 36. Titanium	94
§ 37. Zirconium	96
§ 38. Thorium	97
§ 39. Silicon	99
§ 40. Carbon	101
§ 41. Boron	102
§ 42. Aluminium	103
§ 43. Iron	105
§ 44. Chromium	107
§ 45. Vanadium	109
§ 46. Niobium	110
§ 47. Tantalum	111
§ 48. Bismuth	113
§ 49. Antimony	115
§ 50 and § 51. Arsenic—	
(1) *Arsenious Oxide*	117
(2) *Arsenic Acid*	118
§ 52. Phosphorus	119

CONTENTS

		PAGE
§ 53. Nitrogen	.	120
§ 54. Sulphur	.	123
§ 55. Selenium	.	124
§ 56. Tellurium	.	125
§ 57. Molybdenum	.	127
§ 58. Tungsten	.	128
§ 59. Uranium	.	130
§ 60. Chlorine	.	131
§ 61. Bromine	.	132
§ 62. Iodine	.	133
§ 63. Fluorine	.	135
VI. TABLE OF REACTIONS	.	137

PART II

APPLICATION TO THE ANALYTICAL EXAMINATION OF MIXED COMPOUNDS

1. SYSTEMATIC SCHEME OF EXAMINATION

(1) PRELIMINARY TESTS

A. *Liquid Substances*

§ 64. Testing for Reaction	143
§ 65. Testing for Volatile Substances	144
§ 66. Evaporation Test	146

B. *Solid Substances*

§ 67. Solution	148
§ 68. Sublimation Tests	149
§ 69. Sublimation of Oxidation Products	149
§ 70. Sublimation of Chlorides	151
§ 71. Sublimation of Water	153

(2) EXAMINATION IN THE WET WAY

§ 72. Given an Aqueous Solution	154

	PAGE
§ 73. Given a Solution in Nitric Acid—Precipitation of Oxides and Basic Nitrates	155
§ 74. Precipitation of Chlorides and Iodides	156
§ 75. Precipitation of Carbonates	157
§ 76. Precipitation of Oxalates	158
§ 77. Separation of Alkali Metals from Magnesium	158
§ 78. Examination for Acids	159
§ 79. Elimination of Phosphoric and Arsenic Acid	160
§ 80. Separation of the Alkali Metals (K, Na, Li, Rb, Cs, Tl)	161
§ 81. Solutions containing Hydrochloric Acid	163
§ 82. Analytical Examination of Sulphates	164
§ 83. A Method for distinguishing the Sulphates of the Barium Group (Ba, Sr, Pb, Ca)	165
§ 84. Treatment of the Mixed Sulphates of Bismuth, Calcium, and Sodium	166

II. APPLICATION OF MICROCHEMICAL ANALYSIS TO EXAMINATION OF WATER

§ 85. Qualitative Water Analysis	167

III. EXAMINATION OF ORES—TRACING OF PRECIOUS METALS

§ 86. Ores with Sulphur, Arsenic, etc.	170
§ 87. Tracing of Precious Metals	171

IV. MICROCHEMICAL EXAMINATION OF ROCKS

(1) EXAMINATION OF SLIDES

§ 88. Cleaning	173
§ 89. Testing for Hard Minerals	174
§ 90. Etching of Polished Specimens	175
§ 91. Testing for Carbonates	176
§ 92. Staining of Etched Specimens	176
§ 93. Methods of Staining	177
§ 94. Result of Staining on Minerals	177
§ 95. Result of Staining on the Ground-mass of Rocks	179

		PAGE
§ 96. Testing for Phosphoric Acid	.	180
§ 97. Testing for Potassium and Aluminium	.	180
§ 98. Disintegration of Rocks	.	181
§ 99. Action of Hydrofluoric Acid on Rocks	.	182
§ 100. Isolating Felspars in Thin Sections	.	183

(2) Examination of Powdered Rocks

§ 101. Testing for Hard Minerals	.	184
§ 102. Extraction with Water	.	184
§ 103. Extraction with Hydrochloric Acid	.	185
§ 104. Method of Extraction	.	186
§ 105. Examination of the Solution in Hydrochloric Acid		186
§ 106. Interpretation of Results—Holocrystalline Rocks		187
§ 107. Hemicrystalline Rocks	.	188
§ 108. Fractional Decomposition with Hydrofluoric Acid		189
§ 109. Action of Hydrofluoric Acid on the Rock-forming Minerals	.	190
§ 110. Action of Hydrofluoric Acid on the Accessory Minerals of Rocks	.	191

V. Examination of Alloys

(A) General Remarks

§ 111. Preparation of Specimens	.	192
§ 112. Examination of Hardness	.	193
§ 113. Colouring of Specimens by Heating	.	193
§ 114. Etching of Specimens	.	195
§ 115. Specimens for Fractional Examination	.	197

(B) Details of Microchemical Examination

(1) *Iron*

§ 116. Carbon in Iron	.	198
§ 117. Silicon in Iron	.	198
§ 118. Phosphorus in Iron		199
§ 119. Sulphur in Iron	.	199

		PAGE
§ 120. Manganese in Iron	.	200
§ 121. Chromium in Iron	.	201
§ 122. Tungsten in Iron	.	202
§ 123. Aluminium in Iron	.	203
§ 124. Nickel and Copper in Iron	.	203

(2) *Copper and its Alloys*

§ 125. Cuprous Oxide in Copper	.	204
§ 126. Sulphur, Phosphorus, and Arsenic in Copper	.	204
§ 127. Antimony, Bismuth, and Lead in Copper	.	204
§ 128. Alloys of Copper—Copper and Tin, Bronze	.	205
§ 129. Impurities in Bronze	.	206
§ 130. Copper and Zinc, Common Brass	.	207
§ 131. Impurities in Brass	.	207
§ 132. Statuary Bronze	.	208
§ 133. Copper and Aluminium, Aluminium Bronze and Aluminium Brass	.	209
§ 134. Copper with small Percentage of Aluminium	.	209
§ 135. Copper and Silicon—Silicon Bronze and Cowles's Bronzes	.	210
§ 136. Effect of Silicon on Copper Alloys	.	210
§ 137. Testing of Copper Alloys containing Silicon	.	211
§ 138. Copper with Silicon and Aluminium	.	211
§ 139. Copper and Manganese—Manganese Bronze and Manganese Brass	.	212
§ 140. Copper and Nickel—Nickeline, Manganine, German Silver	.	213

(3) *Alloys of Lead, Tin, and other Metals*

§ 141. Lead, Tin, and Antimony	.	214
§ 142. Alloys for *Clichés*	.	216
§ 143. Alloys for Bearings, Antifriction Metals	.	217

(4) *Alloys of the Precious Metals*

§ 144. General Properties	.	219
§ 145. Gold Alloys	.	220

	PAGE
§ 146. Silver Alloys	221
§ 147. Platinum Alloys	222

VI. EXAMINATION OF SOME COMBINATIONS OF RARE ELEMENTS

§ 148. Native Platinum (Platinum Ore)	222
§ 149. Osmic Acid	224
§ 150. Examination of a Solution containing Platinum, Iridium, Palladium, and Rhodium	226
§ 151. Minerals containing Tantalic and Niobic Acids	228
§ 152. Minerals containing Thorium, Titanium, Zirconium, Yttrium, etc.	231
§ 153. Titanates and Zirconates—Vanadium	233
§ 154. Tin and Tungsten	235
§ 155. Compounds of the Cerite Metals, associated with Compounds of Iron, Aluminium, and Bivalent Metals	237
§ 156. Chromium and Aluminium	238
§ 157. Aluminium and Beryllium	238
§ 158. Beryllium and Magnesium	240
§ 159. Beryllium, Iron, and Manganese—Beryllium and Zinc	241
§ 160. Beryllium, Aluminium, Iron, Yttrium, and Calcium	241

INDEX . 243

INTRODUCTION

MODERN methods of petrographical research—especially those carried out by the aid of the microscope upon thin sections and crushed fragments—have rendered comparatively easy the task of discriminating the mineral constituents of rocks, by the study of their several physical properties, and especially of their optical characters.[1] But in spite of the certainty of the results that can in most instances be obtained by these methods, there constantly arise cases in which it is desirable to confirm or supplement the conclusions, thus arrived at, by *chemical tests* applied to the crystals or fragments of minerals of which rocks are built up.

There are two different methods by which the chemical examination of the constituents of rocks may be accomplished :—

I. By employing one or more of the various methods of isolation (which have been brought to a great degree of perfection in recent years) we may separate from the powdered rock a sufficient quantity of each of the individual minerals composing it, to furnish the materials for

[1] A full description of these methods will be found in Rosenbusch's "Microscopical Physiography of the Rock-making Minerals," translated by Professor J. P. Iddings. (Macmillan and Co., 1888.)

a more or less complete chemical analysis—either qualitative or quantitative.

II. Processes have been devised ("microchemical methods") by which minute particles of the rock-forming minerals may be subjected to a more or less complete chemical analysis. These methods of microchemical analysis are either dry ones—which may be either qualitative or quantitative,—or wet ones—which up to the present time have been almost entirely restricted to qualitative determinations.

It is to the discussion of these qualitative microchemical methods, carried on in the wet way, that the following work is mainly devoted, but it may be of advantage at the outset to furnish a short sketch of all the methods at the disposal of the petrologist, when he finds it desirable to investigate the chemical characters of the mineral constituents of a rock. References will also be given to the works in which the details of these methods may be found more fully described.

I. When a rock has been obtained in a powdered condition, either by artificial crushing or by taking advantage of the natural agents of disintegration (as in sands, muds, etc.), it is usually desirable to subject the material to judicious sifting and washing before attempting to separate its several constituents. The chief processes which have been devised for the work of isolation are as follows:—

(1) By the aid of a lens, or the use of a low power of the microscope, it is possible to recognise the several constituent minerals by the differences of colour, lustre, and form which they present, and to pick them out from the mass by the aid of a wetted bristle or camel-hair brush. This method,

which is of course a very laborious one, has been pursued with success by Professor N. S. Maskeleyne and the late Dr. Walter Flight in their studies of meteorites; and also by Dr. F. Heddle in preparing for chemical analysis minerals which only occur in minute crystals intermixed with other materials. By arranging the grains of the powdered rock in grooves cut in a glass slide, and by other mechanical devices, the operation of sorting may be rendered somewhat less tedious and difficult.[1]

(2) When the material consists of angular or flattened and sub-angular or rounded grains (as is the case with many sands and some powdered rocks), a separation may often be effected by allowing the powder to roll down through tubes of paper of different degrees of roughness of surface, or through moistened glass tubes placed at various degrees of inclination. My friend and former pupil, Mr. C. Carus-Wilson, F.G.S., has been able to make some very perfect separations of sands by ingenious applications of this method.[2]

(3) By washing the powdered rock in water, it is possible to separate its several constituents according to their specific gravities. This was the method employed in 1816 by Cordier, in his famous isolation of the minerals in basalt. Professor O. Derby, of Brazil, has lately advocated the use of the miner's "batea" for such washings, and has shown what valuable results may be obtained by its use.[3]

(4) Where some of the constituents of the disintegrated rock consist of flat plates (micas, etc.) or fine rods (rutile needles, etc.), judicious lixiviation of the powdered or muddy material, and the decantation of the supernatant

---

[1] Rosenbusch, Mikroscop. Physiog., 3rd ed., i. p. 253 (Iddings's translation, p. 107).

[2] Ibid., p. 250 (Iddings's translation, p. 106).

[3] Proc. Rochester Acad. Sc., i. (1891), pp. 198-206.

liquid will often enable us to obtain such minute bodies in a state of tolerable purity.[1]

(5) In some cases, as shown by Thoulet, a current of air may be employed with advantage to winnow out the flattened or elongated particles (micas, etc.) from the more rounded fragments with which they are mixed.[2]

(6) In all cases where the particles to be separated have different densities, great advantage is gained by the use of liquids of high specific gravity. The first of these heavy liquids brought into use for the separation of rock constituents was the solution of the potassium and mercuric iodide (with a maximum density of 3.196). This solution was discovered by Sonstadt in 1874,[3] and the employment of it for isolation processes was advocated by Professor A. H. Church in 1877,[4] and afterwards by Thoulet,[5] Goldschmidt,[6] and others. In 1881 Dr. D. Klein proposed the employment of the cadmium borotungstate, either in solution (giving a liquid of density 3.28) or fused at $75°$ C. (with a density of 3.6).[7] In 1883 C. Rohrbach proposed to use the somewhat unstable solution of the barium and mercuric iodide (with a density of 3.588).[8] Brauns in 1888 suggested the employment of the

---

[1] Thürach. Verhandl. Phys. Med. Gesellsch., Wurtzburg, N.F., xviii. (1884), pp. 1, 2. See also Teall, Mineralogical Magazine, vii. (1887), p. 201.

[2] Bull. Soc. Min., France, iii. (1880), pp. 100, 101.

[3] Chem. News, xxix. (1874), pp. 127, 128.

[4] Min. Mag. i. (1877), pp. 237, 238.

[5] Comptes Rendus, lxxxvi. (1878), pp. 454-456.

[6] Neues. Jahrb. f. Min., etc., Beilag. Band i. (1881), p. 179.

[7] Comptes Rendus, xciii. (1881), pp. 318-321. My friend and former student, Mr. W. B. D. Edwards, now of the Geological Survey of India, has carefully investigated the best and cheapest methods of preparing this very valuable reagent (see Geol. Mag., 1891, pp. 273-275).

[8] Neues. Jahrb. f. Min., etc., 1883, p. 186.

very convenient methylene iodide (specific gravity 3.32).[1] Retgers has further shown how the density of several of the heavy liquids may be increased, though generally at the expense of their stability and transparency.[2] On the other hand, R. Bréon has pointed out that where the employment of high temperatures is admissible (200°-400° C.), a series of liquids ranging in density from 2.4 to 5 may be obtained by fusing various mixtures of the zinc and lead chlorides.[3] For the work of separation with these heavy liquids, various ingenious pieces of apparatus have been devised by Church (1877), Harada (1881), Thoulet (1878), Brögger (1884), Smeeth (1888), Sollas (1891), Wülfing (1891), and Evans (1891). And for the determination of the density of the liquid, Cohen's small balance, the spring balance of Joly, and the natural indicators suggested by Goldschmidt, or the artificial ones devised by Hobbs, may be employed.

(7) The separation of those minerals exhibiting magnetic properties from others with which they are mixed, by using an ordinary magnet is a method that has long been employed. In this way magnetite, pyrrhotite, and the natural alloys of iron and nickel may easily be separated from a powdered rock or sand. Fouqué has shown, moreover, that by a powerful electromagnet it is possible to separate the ferromagnesian silicates from the quartz, alumino-alkaline silicates, and other minerals, which do not contain iron in combination; while Rosenbusch has devised a form of armature for the electromagnet, by means of which the separation is facilitated.[4]

[1] Neues. Jahrb. f. Min., etc., 1886 (ii.), pp. 72-78.
[2] Ibid., 1889 (ii.), pp. 185-192.
[3] Bull. Soc. Min., France, iii, (1880), pp. 46-56.
[4] Mémoires présentés par divers savants à l'Académie des Sciences (1874), xxii., No. 11; Mikros. Phys. (3rd ed.), pp. 250-253 (Iddings's translation, p. 107).

(8) Chemical methods of separation have frequently been employed with great success. Weak or dilute acids serve to remove the calcium carbonate from silicates, and boiling acid will separate the magnesium carbonate. By the use of strong hydrochloric acid, basic and hydrous silicates may be decomposed, and the normal and anhydrous ones left behind. Students of meteorites have long employed this method for separating olivine from enstatite. By the use of hydrofluoric acid either as a gas or in solution, most of the silicates may be decomposed; and in some cases hydrofluosilicic acid or ammonium fluoride may be employed instead of the more powerful but dangerous acid. Mixtures of hydrofluoric with hydrochloric or sulphuric acids decompose nearly all silicates, and in this way rutile, zircon, tourmaline, spinel, andalusite, sillimanite or kyanite, which are not so decomposed, may be isolated from the rocks in which they occur. By treating powders with alkaline solutions, or by fusion with alkaline carbonates, certain constituents may be decomposed or dissolved while others are left intact. Fouqué has shown that by the judicious use of hydrofluoric acid on a powdered rock, first the glassy ground-mass, next the felspars and similar constituents, then the quartz, and finally the ferro-magnesian silicates, are attacked. In this way it has been found possible to isolate from rocks the various porphyritic crystals, preserving their forms and even their brilliant crystalline faces. If the process be skilfully carried out, the most admirable results may be obtained by this method.[1] There are other chemical methods which are possible in certain cases which will readily suggest themselves to the petrographical investigator. A very valuable synopsis of all the separation

[1] Mémoires présentés par divers savants à l'Académie des Sciences (1874), xxii., No. 11.

methods described in special petrographical memoirs down to the year 1890 has been compiled by Professor E. Cohen.[1]

II. The microchemical methods of investigating minute particles of rocks or minerals demand for their successful accomplishment some skill in manipulation and considerable practice; but the results are so rapidly obtained, and are frequently of such extreme value, that no petrologist who has taken the trouble to master the methods will fail to make constant use of them.

The fragments to be employed in a microchemical assay may be obtained by several different methods. When the isolation processes, described in preceding paragraphs, have been employed, one of the particles may be selected for being subjected to the microchemical tests as a preliminary to their complete quantitative analysis. In other cases a convenient fragment may usually be obtained by crushing a small portion of the rock and searching for a suitable crystal or piece of a mineral under a lens. Where the mineral to be examined is only sparsely scattered through the rock, however, one of two methods must be employed. A thin transparent slice, ground in the ordinary way, must be detached from the glass to which it is attached either by warming or the use of a solvent, and, by the aid of needle-points, broken up under the low power of a microscope, so as to detach a fragment of the particular mineral which is to be subjected to microchemical examination. The other method consists in adjusting over the mounted slice of rock a cover-glass perforated with a small hole, so that

---

[1] "Zusammenstellung petrographischer Untersuchungsmethoden nebst Angabe der Literatur." Mittheilungen aus dem naturwissenschaftlichen Verein für Neu-Vorpommern und Rügen in Greifswald.

all the section remains covered except the mineral which has to be tested.¹ To the isolated fragment one or more of the following methods of testing may be applied.

(1) Taking into consideration the dry methods of analysis, the first place must be given to the ordinary processes of assaying by the blowpipe. The results given with minerals of the heavy metals are of the most satisfactory kind, and are obtained with the greatest facility; and further, as Plattner and others have shown, it is possible to obtain quantitative as well as qualitative results. In distinguishing between the ordinary rock-forming minerals, however, ordinary blowpipe tests have only a somewhat restricted application.

(2) In 1867 Gustav Rose, and in 1869 Mr. Sorby, proposed to employ the study of the crystals formed in saturated blowpipe beads, in which crystallisation has been set up, as a method of microscopical analysis; and in some cases this method has yielded very interesting and valuable results.²

(3) In 1865 Bunsen devised his method of studying the flame-reactions of minerals; and in 1873 Szabó suggested that when a fragment of a felspar or other silicate is introduced into a Bunsen flame under certain regulated conditions, the degree of coloration of the flame, and the fusibility of the mineral, may be used as means for estimating the proportions of potassium, sodium, or lime

---

[1] To prepare the perforated cover-glasses they should be evenly covered on both sides with a thin coating of wax, and a small spot of the covering removed by the heated end of a round-pointed wire. A drop of hydrofluoric solution being placed on the spot, a hole is soon eaten through the glass.

[2] Monatsber. d. k. Akad. d. Wissensch., Berlin, 1867, pp. 129-147, 450-464; Monthly Microscopical Journal, 1869, pp. 347-352.

present in it. This method, when due precautions are observed, yields very valuable results.[1]

(4) The determination of the fusing points of minerals as a means for their identification has been employed not only by Szabó, but also by Professor Dœlter and Dr. Oebbeke. The invention of the ingenious "Meldometer," by Mr. Joly, promises to afford geologists the means of making direct and rapid comparisons of the fusion points of different substances.[2]

(5) It is by various wet methods, however, that the microchemical analysis of the rock-forming minerals is now chiefly carried on; the recognition of the various precipitates produced during the reactions being effected by the study of their crystals under the microscope.

Although less than twenty years have elapsed since this method was first suggested by the eminent Bohemian mineralogist, Dr. Emmanuel Bořický, there is already a very extensive literature dealing with microchemical processes. In the year 1875 I had the opportunity, during a visit to Prague, of making the acquaintance of the distinguished investigator to whom we are indebted for the inception of the method, and learnt from his own lips the

[1] In the year 1875 I visited Buda-Pesth and obtained from Professor Szabó full instructions in the employment of this method, which he had then just perfected (Verhandl. Geol. Reicht. 1873, pp. 185-192). In the following year, at Professor Szabó's request, I read a paper and exhibited the method in operation at the Special Loan Collection of Scientific Apparatus at South Kensington ("Science Conferences," 1876, Chemistry, Biology, etc., pp. 418, 419). In 1878 Mr. H. T. Burls, F.G.S., made a large number of important determinations by the method in my Geological Laboratory, and at subsequent dates different demonstrators and assistants have employed the method there with great advantage. Mr. (now Professor) G. A. J. Cole, F.G.S., has suggested the use of a simple piece of apparatus to facilitate the operations (see Geol. Mag., 1888, pp. 314, 315).

[2] Proc. Roy. Irish Acad. 3rd ser., ii. (1891), pp. 38-64.

nature of the interesting studies in which he was engaged; but it was not until the year 1877 that his monograph upon the subject actually appeared.[1] From 1875 down to the date of his early and lamented death I was in frequent correspondence with him upon the subject. It may be added that Bořický's methods, as well as those subsequently devised by Behrens, Streng, Haushofer, Renard, and Klément, and others, have been in frequent use in the Geological Laboratories of the Royal College of Science. I am especially indebted to a former assistant, Mr. T. H. Holland, F.G.S., now on the staff of the Geological Survey of India, and Lecturer on Geology in the Presidency College, Calcutta, for very careful investigations carried on in the laboratory with a view to testing and comparing the various microchemical methods which had been devised up to the time of his leaving this country.

With the exception of the recently published translation of Bořický's original memoir, there has not appeared in the English language any manual on microchemical analysis suitable for use in the laboratory, although several such works have been published both in French and German. When, therefore, Professor Behrens asked my co-operation in bringing out an English edition of his book, I did not hesitate to accede to his request, knowing the great need there is of such a work for English and American students, and how complete and thorough is this particular manual.

One of the most serious sources of error in all microchemical methods, as ordinarily employed, arises from the circumstance that slight differences in the conditions under

---

[1] An English translation of this valuable monograph has been published by Professor N. H. Winchell in the Nineteenth Annual Report (1890) of the Geological and Natural History Survey of Minnesota (Minneapolis, 1892).

which crystallisation takes place, or the presence of infinitesimal quantities of foreign materials, may sometimes profoundly modify the forms and appearance of the crop of crystals produced during a reaction. It is to the work of removing such sources of error that Professor Behrens has especially directed his efforts, as will be seen by the notes and cautions appended to his accounts of the various reactions. In addition to many new and elegant methods which we owe to the author of this work, we are especially indebted to him for the rigorous testing of all the older methods, and for the data he has obtained concerning their reliability and the limits of their applicability and delicacy. Geologists will also find much that is new and interesting in the processes which the author of this work has devised for testing the hardness of the minerals in rocks, and of making use of their susceptibility to take polish as a means of recognising the several constituents of the rock while the grinding of the slide is in progress. Now that so many of the rarer elements are being found to be much more widely diffused in the earth's crust than was formerly supposed, Professor Behrens's new microchemical methods for detecting these substances will be hailed by all petrologists as of the greatest value.

It will be seen that the author of this work, in his desire to make the work of Microchemical Analysis easy and expeditious, relies chiefly on the form and general characters of the crops of crystals obtained as a means of recognition, and only in special cases advocates the investigation of their optical properties. Those who desire to carry the latter kind of investigation further, however, will find an excellent guide in the memoir which has recently been published by my friend, General C. A. M'Mahon, F.G.S.[1]

[1] Mineralogical Magazine, x. (1892), pp. 79-122.

The author of this paper follows the same plan as Professor Behrens in converting all the compounds in the assay, where possible, into sulphates; and he shows that in many of these artificial crystals, in spite of their minuteness and variation in size, it is possible by the ordinary petrographical methods to determine their system of crystallisation, the position of their optic axes, their refraction, double refraction, pleochroism, absorption, and other optical properties; and to employ these determinations as a means of discriminating one sulphate from another.

In these introductory remarks I have almost entirely confined myself to the consideration of the geological applications of microchemical analysis. But it is evident that, like blowpipe assaying, the methods described in this work may be often employed with advantage in the ordinary chemical laboratory, either for rapid testing or in confirmation of the results obtained by other processes. Archæologists and metallurgists, too, will find the methods for examining alloys of great service, especially in cases like those of manufactured articles or objects of art, in which only very minute quantities of the material are available for analysis.

In concluding these remarks, it is only right to add that the English translation of this book has been made by the author himself, and that my own task has been limited to the revision of his manuscript and seeing the work through the press. In the latter task I have received much valuable assistance from Mr. A. E. Tutton, A.R.C.S., Demonstrator in the Chemical Division of the Royal College of Science. Mr. Tutton, who has himself devoted much attention to the crystallographical characters of precipitates, has rendered Professor Behrens and myself an invaluable service by so far revising the nomenclature and formulæ as to make them quite familiar to English students. The illustrations drawn by the author are from *clichés*, supplied by the publisher of the French edition, Mdme. Dunod of Paris.

It may be of interest to add that the whole of the different kinds of apparatus referred to in this work, as suitable for the isolation and analysis of the minerals of rocks, are exhibited in the public Science Museum adjoining the Geological Laboratories of the Royal College of Science, South Kensington, and that, on application to the officers in charge of the collections, every facility will be given to students and others who are desirous of examining this apparatus.

JOHN W. JUDD.

GEOLOGICAL LABORATORIES,
 ROYAL COLLEGE OF SCIENCE,
  SOUTH KENSINGTON, LONDON, S.W.
   30th September 1893.

PART I

GENERAL METHOD AND REACTIONS

I. HISTORICAL REMARKS

In 1877 Bořický published a memoir on the chemical analysis of rocks.[1] This short article originated a new application of the microscope for scientific research. A few reagents were introduced into microscopical practice half a century ago, chiefly by botanists; and somewhat later, crystals of some chemical compounds were examined under the microscope by Harting. In the second volume of his work on the microscope (1866), that author described the forms of precipitated carbonate and oxalate of calcium and of sulphate of barium. In botanical and zoological research reagents were generally employed to produce contrasts of tints and phenomena of imbibition in microscopical preparations of vegetable and animal tissues. Imbibition was out of the question in Bořický's investigations, the material on which he had to work being impermeable rock. He had to seek a new method in quite another direction. What he required were chemical

[1] Bořický, Elemente einer neuen chemisch-mikroskopischen Mineral- und Gesteins-Analyse. Archiv d. naturwiss. Landesdurchforsch. von Böhmen, Band iii., Prag, 1877.

tests, suitable for microscopical use, that could be applied to the minute grains of minerals, from which crystalline rocks are built up. The solvent which he employs for decomposing the silicates—hydrofluosilicic acid—is a distinct feature of his method. It serves him for solvent, and at the same time it is his chief reagent. A few other reagents,—sulphuric acid, chlorine, ammonium sulphide,—are occasionally called in, to complete and to control dubious tests. Another new feature of his method is to be found in the ample use of crystallographical characteristics. This principle, of choosing for microscopical tests compounds which will yield well-developed crystals, has been adopted by all workers in this field of research. Five years later, the author of this work showed clearly [1] that, for microscopical tests, compounds must be chosen, endowed with great tendency to crystallise, and with great molecular volume. The silicates are attacked with hydrofluoric acid, the resulting fluosilicates and fluoaluminates are decomposed by heating with sulphuric acid, and the sulphates are examined by known methods. From a theoretical point of view we have here a complication; in practice this method works better than the original one of Bořický, as the conversion into sulphates opens an ample choice of infallible and delicate tests. Both papers are limited to tests for the constituents of the more common minerals, occurring in crystalline rocks.

Professor Streng of Giessen has published several papers on the same subject.[2] He has applied microscopical investigation to compounds of some elements

[1] Behrens, Mikrochemische Methoden. Verslagen en Mededeeling. d. Kon. Akadem. v. Wetensch. te Amsterdam, Natuurkund. Afd., 1882. Reprinted in Ann. de l'École Polyt. de Delft, t. i. 1885.

[2] Streng, in Bericht d. oberhess. Gesellsch. f. Natur. und Heilkunde, Band xxii., xxiv.; and in Neues Jahrb. für Mineral. 1885, i.; 1886, i.

that had not previously been investigated in this way, and he has added several valuable reactions to those already known. He has, moreover, perfected the accessories and manipulations of microchemical work. Microscopical examination by morphological and physical methods is kept in the foreground, and much weight is attached to ascertaining the optical peculiarities and the angles of crystals produced in the course of microchemical operations. Microchemical examination appears as a valuable auxiliary to microscopical observations in specimens of rocks prepared by grinding and polishing, and to blowpipe analysis.

In 1885 Haushofer went a step farther. He called his work [1] a manual for distinguishing several elements, and regarded microscopical reactions as supplementary to ordinary qualitative analysis.

When more closely examined, his work is found to contain microscopical reactions for the majority of the elements, the foundation of a microscopical analysis. Some of the tests given by him are, it is true, of little value; the discrimination of some nearly allied elements, such as zinc and cadmium, cobalt and nickel, leaves much to be desired; but, notwithstanding these shortcomings, the tendency of the work is manifest: it is to bring all elements, even such rare ones as tantalum, yttrium, and thorium, within the range of microscopical analysis.

The treatise of Klément and Rénard [2] (1886) is on a level with the manual of Haushofer. It gives a valuable abstract of the microchemical tests known at that time, accompanied by copious literary references, but the

[1] Haushofer, Mikroskopische Reactionen. Eine Anleitung zur Erkennung verschiedener Elemente unter dem Mikroskop, als Supplement der qualitativen Analyse, München, 1885.

[2] C. Klément et A. Rénard, Réactions microchimiques à cristaux et leur application en analyse qualitative, Bruxelles, 1886.

same defects are found as in the work of Haushofer, with immaterial corrections. The expectation, awakened by the second part of the title, is but scantily satisfied.

In a comprehensive sketch, several short papers ought to have been mentioned, and those of O. Lehmann and of Michel-Lévy and Bourgeois should have been reviewed more fully, but this would have filled thrice the space available for these introductory remarks. I have aimed at giving a summary of the development of microchemical analysis during the last ten years, and at the same time an outline of the task undertaken by the inventors of this new branch of chemistry.

II. AIM OF MICROCHEMICAL ANALYSIS

The memoirs of Haushofer and of Klément and Rénard have been of great value to workers with the microscope, whereas they have not found entrance in chemical laboratories. From this circumstance the conclusion might be drawn, that the idea of Haushofer has no vitality, and that microchemical tests ought to remain, where they originated, in the study of the microscopist.

An improvement of Haushofer's idea may be found in extending its range. I venture to say that, by assiduous study, microchemical analysis will be developed into a system that will rival blowpipe analysis, as regards its rapidity and its unassuming character with regard to space and laboratory appliances, and in many instances surpass blowpipe analysis in the variety and delicacy of its tests. The advantages that accrue from such a method of investigation to chemistry, and all branches of science and industry allied with it, are so great and manifest, that I make bold to call in the co-operation of all those interested in analytical chemistry.

It seems to me that the reserved attitude of chemists towards microchemical analysis may have been caused, in the first place, by the belief that microchemical tests cannot be of any use without a long apprenticeship in the use of the microscope, and that they are therefore apt to lead the observer into errors; and, in the second place, by a conviction of the unsatisfactory nature of many microchemical tests.

The remarks made by O. Lehmann about the little reliance which is to be put on microscopical examination of crystals have an important bearing on microchemical analysis. The measuring of angles under the microscope is, it is true, subject to errors from more than one cause, and I should be one of the last to deny that the investigation of optical peculiarities in microscopical crystals is a delicate task; I should indeed go so far as to say it is a vexatious one. An unusual position or unfavourable illumination of a minute crystal will sometimes reduce the observer to a condition of despair whilst he is trying to make out its crystallographical and optical description. But these objections belong to the morphological department; they lose much of their weight when you make *chemical* tests the leading ones, and put the idea of *microchemical* analysis in the place of analysis aided by microscopical investigation. If microchemical tests had been worked out by chemists instead of by geologists, the form of crystals or the angles of their faces would certainly never have held the first place among the characteristics employed.

If all elements are to be brought within the range of microchemical analysis, the *chemical* behaviour of the objects must recover its legitimate place at the head of the characteristics. A crystalline precipitate, produced by means of platinum chloride, is found to be composed of octahedra. Hereupon the conclusion may be founded, that barium is absent, while it would not be sound reasoning to deduce the presence of potassium—unless it can be made highly improbable by other considerations, that ammonium, rubidium, or cæsium is present.

With an ample choice of characteristic compounds, preference will be given to reactions producing compounds that are readily perceived and recognised. For microchemical tests the larger crystals of the sulphate of calcium will be preferred to the very small crystals of the carbonate and oxalate; for the same reason chromate of silver will rank before the chloride. By this device the last difficulty is removed—namely, the necessity of employing high magnifying powers, requiring much practice and giving at the best but a narrow field of vision. As a general rule, a power of 50 will be found sufficient, giving a field that will take in objects of 3 mm. In a few cases a magnifying power of 200 must be employed. A chemist who has now and then made use of the microscope will train himself in the course of a few weeks, whilst trying the more important tests. Figures and descriptions are of little value as compared with the actual experiment and observation of all that happens in the course of a microchemical reaction. A preliminary trial of the more important reactions is therefore indispensable, and going to work upon microchemical analysis without such training will have indeed very little chance of success.

The collection of reactions given by Klément and Rénard shows several gaps, and in the manual recently published by Professor Streng[1] certain tests are wanting—

[1] Streng, Anleitung zum Bestimmen der Mineralien, Giessen, 1890.

among others means for detecting cadmium and nickel. Other reactions recommended by these authors leave much to be desired. I must content myself with these short remarks, as it is of more importance to establish the principles on which the choice of microchemical reactions depends.

1. *A Minimum Quantity of the Substance.*—This has been the starting-point of microchemical analysis. Reactions of extreme delicacy are required for operating on hundredth parts, or with some elements (Cl, Mg, Pt, Tl) on millionth parts of a milligramme. This delicacy depends on a feeble degree of solubility of the characteristic compounds, on their molecular volume, and at least even as much on their aptness to form large crystals. In the last place it depends on the size of the tested drops—that is to say, on a high degree of concentration of the assay and the reagents employed. For instance, the sulphates of barium and calcium and cæsium alum are compounds that are turned to account for tracing sulphuric acid. Their solubility is expressed by the numbers 1 : 400000 ; 1 : 400 ; 1 : 200. These numbers explain the frequent use of barium chloride in ordinary qualitative analysis and the restricted application of calcium chloride. In microchemical analysis the position is reversed. Here the question is decided by the consideration, that sulphate of barium exhibits grains of extreme minuteness, while sulphate of calcium yields neatly defined, characteristic crystals, easily recognised under a power of 100. Slow evaporation favours the growth and increases the number of crystals. After some practice we are able to trace in this way 0.00019 mgr. (0.19 μgr.) of sulphur. Cæsium alum indicates 0.12 μgr. of sulphur, notwithstanding its higher degree of solubility. This is due to its great molecular volume and to the large

size of its crystals. In sulphate of lead a decided tendency towards crystallisation is found combined with great molecular volume and a low degree of solubility (1 : 23000). In consequence of this it is possible to trace 0.006 μgr. of sulphur.

Silver can be traced under the microscope as chloride, crystallising from its solution in ammonia. It can likewise be discovered as chromate, precipitated from an acid solution. The delicacy is nearly the same for these two tests, while the molecular volume of the chromate is not much greater than that of the chloride, which is nearly insoluble in water. The limit of an unequivocal reaction is for both tests 0.15 μgr. of silver, but this small quantity is divided into a great number of minute crystals, when operating with the chloride. Accordingly, a power of 300 must be employed to render them distinctly visible, while the crystals of the chromate grow to a size for which a power of 50 is amply sufficient. Here the delicacy of the test is increased by the accumulation of matter around a small number of points. Amorphous precipitates are of little value for microchemical tests. The same holds true for coloured liquids, even when the change of colour is very striking. The colour of powdery precipitates and of coloured liquids becomes almost imperceptible under high powers.

The relation between the delicacy of a reaction and the size of the tested drops is manifest. Accordingly, the liquid that is to undergo examination must be concentrated as far as possible, the reagents must be employed in concentrated solutions, or, if possible, as dry powders. Drops of one cubic millimetre have been employed for establishing the limits of the reactions described in this work. Such drops will spread on a slide over a circular space of 2 mm. diameter, comprised in the field of a power of 50.

2. *A Minimum of Time.*—I am not sure if I have put this principle in its proper place, because microchemical analysis cannot come into universal use, if this condition is not attended to in the first place.

Petrographers will of course attach great weight to the possibility of analysing minute particles of fine-grained rocks, but in chemical and industrial laboratories it will be of the greatest moment to *save time* by the application of microchemical methods.[1]

With a view to this object, all reactions necessitating a tedious preparation ought to be excluded or relegated to a subordinate place, and likewise all reactions which require a long time for their accomplishment.

For instance, the method of Streng for detecting beryllium by crystallisation of its chloroplatinate must be rejected, because the crystals will not appear otherwise than in a desiccator. On the other hand, precipitation of zirconium with potassium sulphate is open to objection, since you have to wait half an hour before it becomes visible. For similar reasons the method for separating nickel from cobalt, based upon the behaviour of their oxalates when dissolved in ammonia, which has recently been recommended anew by Haushofer, must be refused a place among microchemical reactions. This method labours under the disadvantage of sometimes requiring over half a day or even longer for its completion.

Filtration must be regarded as an extreme remedy, employed for some valuable tests which require a clear solu-

---

[1] The examination of a solution containing the following elements, Ca, Mg, Zn, Mn, Co, Ni, has been accomplished in forty minutes; another solution containing Ag, Hg, Pb, Bi, Sn, Sb, As, has been examined in an hour. An examination of this kind may be worked out on one slide, in such fashion, that all the characteristic products are arranged next each other—a saving of space, likewise not to be overlooked.

tion absolutely free from suspended matter. Here the advantage of scanty solubility combined with characteristic crystallisation makes itself clearly apparent. To examine a solution containing calcium, magnesium, and aluminium in the usual way, aluminium must be precipitated with ammonium sulphydrate or with caustic ammonia; calcium with ammonium oxalate, and magnesium with sodium phosphate, from an ammoniacal solution.

This method is burdened with two filtrations, and much time is lost in waiting for complete precipitation of the oxalate of calcium; moreover, if at last the solution has become highly diluted, you have to wait once more for the precipitate of ammonium magnesium phosphate. It is not possible to proceed by a shorter way, considering that the flakes of aluminium hydroxide and the lumps of calcium phosphate would be an obstacle to the observation of a microcrystalline precipitate. With microchemical reactions the problem is solved in quite another way. By adding sulphuric acid to a drop of the mixed solutions, characteristic crystals of calcium sulphate ($CaSO_4 + 2H_2O$) are produced; after addition of a cæsium salt the same drop yields large crystals of cæsium alum; finally, addition of ammonium chloride, ammonia and sodium phosphate, will precipitate from the same drop minute crystals of ammonium magnesium phosphate. By working with a hot solution, the forms of these minute crystals can be made so characteristic, that calcium and aluminium must be present in very great excess to conceal them from observation.

3. *Certainty of the Reactions.*—Three conditions must be fulfilled to make a reaction reliable. Its occurrence should not be doubtful; its observation should not be attended with difficulty; the reaction must be characteristic. Reactions which are influenced by small variations of tem-

perature or by slight modifications in the composition of the solvent are of no value. Cadmium cannot be distinguished from zinc by means of oxalic acid, because the solubility of cadmium oxalate has a wide range, depending on the proportion of free acid. Tartrate of barium and antimonyl is not to be recommended for discovering barium or antimony, since its crystallisation is hampered by large quantities of alkaline salts.

Modification of characteristic compounds, caused by the interference of a third or fourth element, is of frequent occurrence. Sometimes it can be turned to profit for new reactions, as, for example, the remarkable influence of gold chloride on thallous chloride; but on the whole it is not favourable to microchemical analysis, masking and spoiling reactions, some of which are rendered useless in compound solutions, whilst they are of high value in simple ones. Thus, the presence of cadmium and of zinc can be demonstrated by means of oxalic acid; if both elements are present in the same drop, only the presence of zinc is established. If zinc is accompanied by a slight admixture of magnesium, oxalic acid will precipitate shapeless crystalline lumps; if magnesium predominates, well-defined hexagonal plates of a double oxalate will separate out, having nothing in common with the rod-like forms of zinc oxalate. Solutions containing equal weights of barium and strontium will yield compound crystals of sulphate, indicating only the presence of strontium. The tartrate and carbonate of calcium form beautiful crystals, whereas an admixture of barium leads to powdery or spherulitic precipitates, quite useless for microchemical analysis. Potassium chloride provokes singular modifications in the crystals of lead chloride; ammonium chloride spoils this reaction entirely. The disturbing influence of ammonium chloride on many reactions has been the subject of elaborate research, whereas very little is known of the disturbing influence of boric acid and soluble borates, by whose presence the characteristics of several tartrates and oxalates are entirely obscured.

To make the reactions readily perceptible, which is desirable with a view to exact and expeditious work, the size of characteristic crystals must not fall below the limit for clear definition under a power of 200. Tests requiring powers ranging from 400 to 600 for discovering minute crystals, concealed in a mass of powdery or fragmentary matter, should be discarded.

III. APPARATUS

The Microscopic and other Appliances.—Any microscope armed with powers ranging from 50 to 200 can be utilised for microchemical work. A set of polarising prisms and fittings for measuring angles are valuable additions. Special fittings for convergent light are superfluous; the same is to be said of immersion lenses. A mechanical stage is quite unfit for chemical manipulations; it would very soon be spoiled and put out of gear. Fittings for heating the objects are very serviceable for biological and petrographical research; for chemical work they are too slow in their action. A *long focus* is of great importance. It permits of the objects being placed in proper position with ease and rapidity; it facilitates the picking out of minute particles and the rapid addition of reagents. In addition to these advantages the chance of spoiling the undermost lens is diminished.

Strong object lenses and feeble eye-pieces are usually employed for microscopical research. For microchemical work preference should be given to the combination of feeble object lenses with strong eye-pieces. For the majority of the observations hereafter described, an object lens has been employed which has a distance of 3 cm. from the object. With medium and strong eye-pieces its power varies from 36 to 90-fold. The diameter of its field can be extended to 3 mm.

Protection of the undermost lens is rendered unnecessary by a long focal distance, excepting in a few cases of peculiar nature. By frequent rinsing and wiping with filter-paper or with clean soft linen it may be kept in good condition for several years. I would have sulphuretted hydrogen, it is true, banished from the room in which microchemical work is going on. I should also like to restrict the use of ammonium sulphydrate, and to have it under the microscope as short a time as possible. Hydrochloric and nitric acids are generally employed in a dilute state and in small quantity. They do not demand special precautions, unless high powers are employed. Hydrofluoric acid and ammonium fluoride are always to be used with caution, even under lenses of long focal distance. A precaution which suggests itself at once consists in covering the object with a small watch glass or a cover glass, resting on two wires or threads of glass.[1] Another expedient is found in protecting the object lens with a small round cover glass, affixed by means of a drop of water or glycerine. It has the advantage of not interfering with the manipulation of the object. Of course the lens must be washed and wiped before setting aside the microscope.

[1] Streng employs large cover glasses resting on small pieces of cork, glued under the corners. Cleaning such covers is a disagreeable task, in consequence of their fragility. Short slides of thin glass can be fitted for covers by rounding their corners in the glass-blower's flame and setting them up about 1.5 mm.

A micrometer scale in one of the eye-pieces is serviceable for rough determination. Its use is mastered after a few trials. To make drawings of microscopic crystals is quite another thing. Where microchemical tests are called in as an auxiliary to ordinary qualitative analysis, a rough sketch, such as can be made without special exercise, will now and then be necessary. On the other hand, when microchemical analysis is employed for scientific research, drawings of new or uncommon forms of crystals may be very helpful, especially when accompanied by a short notice detailing the circumstances under which the new crystals were formed. Photography cannot take the place of sketching, because it is by no means easy to take neat photographs of highly refractive colourless crystals. Besides this, the impartiality of photography reproducing shapeless lumps, partly out of focus, along with well-developed and neatly-defined crystals, often proves very irritating to the observer. A vertical arrangement is better suited for photographic reproduction of microchemical preparations than a horizontal camera. The height of the structure has some inconvenience, but the consideration that photographs can be taken without altering the position of the objects decides the question in favour of the vertical camera.

With the aid of a camera-lucida of simple construction any observer is able to draw a tolerably just outline of microscopic crystals. Practised draughtsmen will avail themselves of the camera-lucida to save time and to be sure of rendering the details in their just size and proportion. They will take off the camera-lucida when finishing the sketch by putting in shades and half tints. In the meantime, it may as well be noted that even for exact reproduction of mere outlines some practice is necessary.

Complicated fittings for illumination are to be rejected.

A good lamp, fitted with condensers for the light (system Kochs and Wolz), will render valuable service in dark foggy days. Spectroscopic observation may be useful (*e.g.* for compounds of didymium and erbium), but the adjustment of apparatus and objects is too tedious for ordinary work.

Some remarks about the testing of a microscope for chemical work will not be misplaced. Messrs. W. and H. Seibert, Wetzlar, Prussia, have furnished cheap microscopes for chemists, satisfying all reasonable demands. These microscopes are fitted with two lenses (Seibert O and III.), and two eye-pieces, 2 and 4, the last one with micrometer scale, with rotating stage for measuring angles and with polarising prisms. Price 150 marks = £7 : 10s. The magnifying power can be varied from 40 to 300. Messrs. R. and J. Beck, 68 Cornhill, London, have undertaken to furnish a good microscope for chemical and petrographical purposes at nearly the same price. It is made on the model of their "Star" microscope (with two objectives and two eye-pieces magnifying 40 to 300 diameters), but fitted up with polarising prisms, micrometer scale, and with a rotating stage, divided into single degrees. For testing the low power (40 to 80) potassium chloroplatinate or calcium sulphate can be used

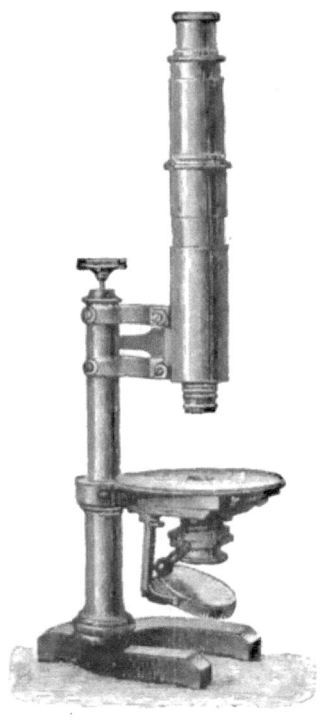

FIG. 1.—Microscope for Chemists (Seibert), $\frac{H}{S}$.

if they are precipitated from moderately diluted solutions, in which crystallisation will set in after a minute. Under these circumstances the octahedra of the chloroplatinate and

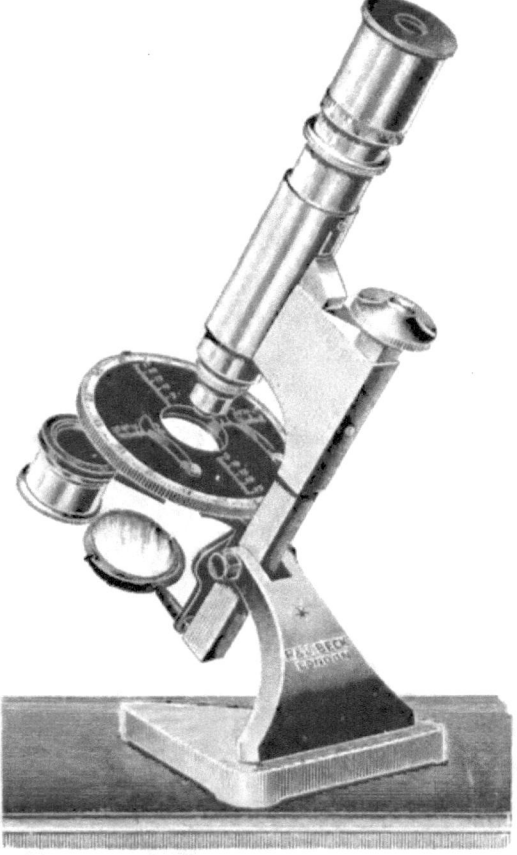

Fig. 2.—Microscope for Chemists (Beck), $\frac{H}{3}$.

the oblique prisms and twins of the calcium sulphate must be readily visible under the lowest power after five minutes. For the higher powers, ammonium-magnesium phosphate is a good test-object. It must be precipitated from a warm ammoniacal solution containing 1 : 2000 of magnesium. Another serviceable test is furnished by the octahedra of silver chloride, crystallising from an ammoniacal solution containing 1 : 2000 of AgCl.

After treating of the microscope, a few words are to be said about the *slides*. They perform manifold service; besides supporting the objects during microscopical investigation, they are put to the same uses as test-tubes and evaporating dishes in ordinary analysis. Slides furnished by Messrs. Seibert of Wetzlar, and by Fuess of Berlin, were found sufficiently thin and well tempered. With careful handling they will stand heating to the boiling-point of sulphuric acid. Now and then a slide will crack, it is therefore prudent to be well stocked and to perform delicate operations over a clean porcelain dish. The ordinary size of slides (78 × 26 mm.) is well fitted for current use. Short slides are procured by halving a batch of long ones. They are more convenient in measuring angles, as they permit full rotation of the stage. A dozen of them should have the angles rounded and turned up (see p. 13, footnote) to fit them for covers in sublimation tests and for protecting preparations from dust and from hasty evaporation.

For tests requiring the use of hydrofluoric acid, of silico-hydrofluoric acid, or of ammonium fluoride, the slides must be *coated with Canada balsam*. Varnishing can be done by spreading a thick layer of balsam (if viscous, diluted with essence of turpentine), warming gently, running off the surplus, and heating the coated slides in a stove or on a slab of iron till a test-piece, cooled in water, will not take an impression from the finger-nail. It must be remembered that heating too far will produce an unpleasant yellow tint, and that heating too long will make the coating liable to crack.

C

This method is simple, but it is not easy to manage if smooth surfaces are required. Drying the coated slides at ordinary temperature is very tedious, it requires about two months. For frequent use a solution of hardened Canada balsam in carbon disulphide is to be recommended. Canada balsam is heated in a shallow evaporating dish, till a sample, cooled in water, shows sufficient hardness. After cooling it is scraped out, crushed and dissolved in carbon disulphide to the consistency of olive oil.[1] A few trials are made on broken slides, with a view to tempering the hardness. This is easily done by adding soft balsam. The solution is run over the glass in the same manner as collodion. After ten minutes the slides are ready for use. In cold weather a slight heating of the slides is useful to prevent condensation of water from the atmosphere during the evaporation of the solvent. Slides that have been varnished with due care resist dilute acids a long time. They are cleaned with cold water and burnished by rubbing with soft linen. By alkaline solutions, dilute ammonia included, they are rapidly spoiled.

For heating with hydrofluoric acid, for calcining and fusing with alkali or with acid potassium sulphate, small *spoons of platinum*, from 9 to 15 mm. wide, are employed. Spoons with a riveted handle are troublesome to clean; they should be stamped from *one* piece of sheet platinum, with a lappet of 5 to 7 mm. for a handle, which is seized with pincers tipped with platinum.[2]

---

[1] Streng recommends ether as solvent. It is apt to absorb water from the atmosphere in the course of evaporation. The coating is by these inclusions rendered turbid and liable to infiltration.

[2] For rapid and easy cleaning of platinum spoons it is a good plan to mould them in a small box filled with plaster of Paris. After desiccation the plaster is varnished with shellac, dissolved in alcohol. This treatment will prevent crumbling and dust.

For calcination and sublimation of small quantities, for heating them with strong sulphuric acid, and for many other operations, a strip of sheet platinum, 5 mm. broad and 15 mm. long, made into a *spatula*, by riveting and soldering it between the ends of two strips of nickel plate, will be found very serviceable.

Very small samples are calcined and fused on *platinum wires*, bent in the same fashion as for blowpipe tests. Cleaning of spoons and wires must be done with the utmost care. Heating with strong nitric acid in a porcelain crucible will remove many incrustations. If this remedy fails, fusing with acid potassium sulphate must be resorted to, followed by heating with hydrochloric acid.

Burners for microchemical use must allow a reduction of the flame to a length of 5 mm. For heating and concentrating small drops a flame of 10 mm. is too large. For ordinary work a Bunsen burner may be adapted by unscrewing the tube in which the gas is mixed with air. Where gas is not to be had, common oil, with a thick thread of cotton for a wick, will answer the purpose. It is better than spirit, which will not burn at all or with a flame far too large for microchemical work. A *water bath* of small dimensions[1] will render good service for concentrating and evaporating large drops; for delicate work a naked flame, made as small as possible, is to be preferred.

The *dropping bottles* of commerce are usually adjusted for drops of 0.05 cub. centim., quite unfit for delicate tests. It is to be noted that the experiments by which the limits for the various reactions have been determined have been conducted with drops of 0.001 cub. centim. The orifices of the commercial tubes must be contracted by fusing, and for delicate tests pipettes must be used, drawn from narrow

---

[1] Streng, Anleit. z. Bestimm. d. Mineral. p. 64.

tubes, which will give off drops of 2 to 3 mgr. if the slide is tapped with the point of the pipette, the upper end being closed with the finger. For taking up and for transferring small drops, *capillary tubes* are very serviceable. They have the advantage of being made in so short a time that they may be thrown away after being used, dispensing with tedious and uncertain cleaning.

Reagents are taken up and distributed with *short platinum wires* of 0.5 mm. diameter, mounted in cylindrical handles of glass or wood. If they are bent to small hooks they will take a definite quantity of liquid, provided that they be *perfectly clean*, and this is a weak point. I have, therefore, discarded the hooks in favour of straight wires, which can be cleaned in a few seconds by rinsing, rubbing, and igniting.

Funnels are unfit for microchemical filtrations, even when attached to an air-pump.[1] Only one device for *microchemical filtration* does answer all reasonable requirements—that of Streng.[2] He places the slide bearing the liquid which has to be filtered on a box turned upside down and slightly inclined, and beside this box a clean slide. The filtration is effected by means of a strip of filtering-paper, 2 mm. broad, touching the turbid drop and hanging down on the clean slide. A modification of this device, pointed out by the same author, will be found less simple but more effective. The filtering-paper is cut in the form of a Y, two of the branches are put on a clean slide and affixed to it with a drop of water, the third branch is bent downwards till it touches another slide placed beside the first one, but somewhat lower; finally, the liquid that is to be filtered is put on the upper slide,

---

[1] Recommended by Rosenbusch, Mikrosk. Physiogr. d. Miner. i. 105.
[2] Streng. Anl. z. Bestimm. d. Mineral. p. 65.

in contact with the forked end of the strip of paper. The measures given by Streng—2 mm. breadth and 25 mm. length—may be reduced to 1 mm. and 10 mm. A strip of wood, 7 mm. thick, bevelled to 5 mm. (length 50 mm., breadth 30 mm.), and nailed on a thin piece of board, makes a good support. Under the most favourable circumstances about 5 mgr. of water are retained in the paper; thus the volume of the drops that have to be filtered must be at least 0.01 cub. centim. if dilution is to be avoided.

Of other accessories some may be mentioned here: A pair of forceps tipped with platinum, as used for blowpipe tests; a small agate mortar; strips of tough paper for covering grains of mineral while crushing them in the mortar; clean filtering-paper cut to small 16° size, for absorbing small quantities of liquids, for cleaning platinum wires, etc. Some other accessories, of occasional use, will be described in Part II.

IV. REAGENTS

1. *Distilled Water.*—It is kept in a small phial or dropping bottle, and frequently renewed, because it will take up traces of alkali and silica even from a phial made of hard glass.

2. *Alcohol.*—It is seldom used as solvent. In some cases it is useful for accelerating crystallisation, as, for instance, when dealing with very small quantities of calcium sulphate or of sodium antimonate. It decomposes potassium manganate and reduces potassium osmate to osmite.

3. *Sulphuric Acid.*—A mixture of equal volumes of concentrated acid and water.

4. *Nitric Acid.*—A mixture of equal volumes of water and of nitric acid of sp. gr. 1.4.

5. *Hydrochloric Acid.*—The acid commonly used for qualitative analysis, of sp. gr. 1.12. The acids 3, 4, 5 must from time to time be tested for calcium, potassium, and ammonium. Some drops are evaporated on platinum; if any considerable residue is left, the acid must be thrown away. A volatile residue points to ammonium. In this case something is amiss with the stoppers.

6. *Hydrofluoric Acid.*—Pure acid, leaving no residue of fluor-salts when evaporated on platinum. It is kept in bottles of ebonite; small quantities in stoppered tubes of the same material.[1] As the keeping of this reagent is often troublesome, preference will generally be given to the following one :—

7. *Ammonium Fluoride.*—It is tested for silicon with sodium chloride (§ 39, *a*). Purification is effected by heating with a little ammonium hydrate and by subliming the dry mass in a platinum crucible. The first and the last quarter of the sublimate is thrown away. It is crushed to coarse powder and kept in a tube of ebonite. Its solution in hydrochloric or in sulphuric acid is employed in the same way as hydrofluoric acid. Its action is, however, less violent. It is necessary to heat strongly after evaporation of the solvent, to make sure of the expulsion of all ammoniacal compounds; otherwise erroneous conclusions may be drawn as to the presence of potassium, rubidium, or cæsium.

8. *Acetic Acid.*—Glacial acid, diluted with one-tenth part of water. Used for neutralising and for acidulating, also for retarding some reactions.

8B. *Formic Acid* is sometimes used for precipitating the cerite metals.

---

[1] Such tubes for the reagents 6 and 7 are furnished by Dr. R. Muencke, Berlin, Luisenstrasse 58.

9. *Oxalic Acid.*—Small crystals or powder. A precipitant of frequent use.

10. *Tartaric Acid*, coarsely powdered. Sometimes used for dissolving antimonous oxide.

11. *Silica.*—The fine powder resulting from the preparation of hydrofluosilicic acid.

12. *Caustic Potash.*— ⎫ These reagents are crushed to
13. *Caustic Soda.*— ⎭ coarse powder while hot. Solutions must not be kept in glass bottles, since they will take up silica. Soda is employed for precipitating niobic and tantalic acids, and for dissolving and tracing zinc oxide.

14. *Caustic Ammonia.*—The liquid employed in qualitative analysis. It should not be kept in the same box with nitric and hydrochloric acids, as crusts of ammonium salts would form around the stoppers.

15. *Magnesium.*—The powdered metal of commerce. Used for reduction of nitrates and of selenium dioxide.

16. *Zinc.*—A small rod of pure zinc is heated to $150°$ C., when it can easily be flattened by hammering into thin laminæ. Used for reduction of several metals.

16B. *Iron.*—Thin sheet iron, brightened by filing or by rubbing with glass paper. Used for precipitating copper, and for extemporising solutions of ferrous chloride.

16C. *Tin-foil and Sheet Tin.*—Used for reduction of antimony and bismuth, for precipitating phosphoric and arsenic acids, and for producing stannous chloride.

17. *Potassium Nitrate.*— ⎫ Small crystals or powder.
18. *Potassium Chlorate.*— ⎭

19. *Hydrogen Peroxide.*—Sometimes used as an oxidising agent.

20. *Potassium Nitrite.*—Kept as saturated solution. Used for oxidations and for precipitation of cobalt, nickel, copper, lead, and rhodium.

21. *Sodium Bicarbonate.*—Crystalline powder. As a precipitant, it is generally superior to the ordinary carbonate. If the latter must be employed, it is extemporised by heating a small quantity of bicarbonate.

22. *Ammonium Carbonate.*—The carbonate of commerce, powdered. It is employed for precipitations in the same way as the sodium compound, and for dissolving carbonates and oxalates of beryllium, uranium, yttrium, and thorium. To prevent volatilisation, it is a good practice to grease the stopper with a trace of vaseline.

23. *Rubidium Chloride*, powdered.—It is used in the place of potassium chloride for tracing platinum, iridium, titanium, and zirconium, being a more delicate test.

24. *Cæsium Chloride.*—Kept and used like rubidium chloride, which it surpasses in delicacy. It is the best of all reagents for detecting aluminium.

25. *Sodium Chloride*, powdered.—Employed for detecting silicon, fluorine, and antimonic acid.

26. *Ammonium Chloride*, powdered.—It is used in tracing magnesium, vanadium, and platinum.

27. *Potassium Iodide* is used for tracing lead, thallium, palladium, mercury, antimony, bismuth, arsenic, and selenium. For separations, the *ammonium* compound is to be preferred, although it is more troublesome to keep, being deliquescent and slowly decomposed by the oxygen of the atmosphere.

28. *Potassium Ferrocyanide.*—Stove-dried and reduced to powder. Besides the well-known colour tests in ferric and cupric solutions, it produces crystalline precipitates with soluble compounds of cerium, didymium, barium, and calcium. The *ferricyanide* is seldom used. It yields characteristic precipitates with zinc and cadmium.

29. *Ammonium Thiocyanate.* — A saturated solution. Added to cobaltous acetate, it is employed in testing for mercury; added to mercuric chloride until the precipitate is redissolved, it affords a valuable test for cobalt, copper, and zinc.

30. *Ammonium Silicofluoride.*—It is prepared by saturating hydrofluosilicic acid with ammonium carbonate, or by heating a strong solution of ammonium fluoride with an excess of silica till the smell of ammonia is no more perceived. The resulting liquor is evaporated and the residue is subjected to sublimation. The first quarter may contain ammonium fluoride. This reagent is used in the place of hydrofluosilicic acid for detecting sodium and barium.

31. *Potassium Sulphate*, coarse powder.—Used for detecting bismuth. Mixed with sulphuric acid it is employed for decomposition in the dry way.

32. *Sodium Sulphate* is employed for precipitating the cerium metals, separating them from yttrium and zirconium.

33. *Potassium Bichromate.*—Of great value for detecting silver and for distinguishing barium from strontium. *Ammonium bichromate* keeps as well and is to be preferred with a view to separations.

34. *Potassium Oxalate.*—Reagent for beryllium and tin.

35. *Acid Oxalate of Potassium.*—Reagent for bismuth and for zirconium. Both oxalates must be tested for impurities. A few small crystals are put in a drop of water under a power of 80. If minute crystals are thrown off, which remain a long time undissolved, the reagent must be rejected. Purifying by recrystallisation would be too tedious.

36. *Potassium Tartrate*, or, in its place, *Seignette Salt*, stove-dried and powdered.—It precipitates splendid crystals from solutions of calcium and strontium.

37. *Sodium Acetate*, dried and powdered.—Sometimes used as a reagent for uranium and for tantalic and niobic acids. Frequently employed for mitigating the effect of free strong acids. For the latter purpose a saturated solution of *ammonium acetate* is to be preferred.

38. *Sodium Phosphate*, dried and powdered.—Precipitates magnesium from ammoniacal, molybdic and tungstic acids from acid solutions.

39. *Ammonium Molybdate*, powdered.—Its solution in weak nitric acid should leave a white fine-grained residue. Yellow crystalline grains betray the presence of phosphorus or silicon.

40. *Thallous Nitrate*, powdered.—One of the most valuable reagents for microchemical analysis. It is employed in testing for the halogens, for gold, for chromic, vanadic, molybdic, and tungstic acids; for yttrium, uranyl, and thorium. The sulphate is also very serviceable, but for separations the nitrate has an advantage.

41. *Barium Acetate*, powdered.—Reagent for silicon and for fluorine.

42. *Strontium Acetate*, fine powder.—Reagent for carbonic acid.

43. *Calcium Acetate*, powdered.—A valuable reagent for tracing sulphuric acid and arsenic acid.

44. *Magnesium Acetate*, saturated solution.—It is utilised in detecting sodium and uranyl, and in testing for phosphoric acid.

45. *Zinc Acetate*, powdered.—It is helpful in testing for cobalt and copper. In testing for arsenic acid it can be used in the place of calcium acetate, in testing for sodium and uranyl it can be employed instead of magnesium acetate, if the latter should be found to contain traces of sodium.

46. *Lead Acetate*, powdered.—Employed in tests for copper, nickel, and for sulphuric and chromic acids.

47. *Cupric Acetate*, powdered.—Utilised for detecting lead, zinc, and platinum.

48. *Cobalt Acetate*, powdered.—It is employed in testing for mercury.

49. *Uranyl Acetate*, powdered.—Affords a very characteristic test for sodium.

50. *Bismuth Nitrate.*—The basic nitrate of commerce gives no trouble in keeping. Its solution in sulphuric acid is employed as a reagent for potassium and sodium.

51. *Silver Nitrate*, powdered.—Reagent for chromic and vanadic acids.

52. *Mercuric Chloride*, powdered.—Employed for detecting stannous compounds and ammonium. Combined with ammonium thiocyanate it affords characteristic tests for cobalt, copper, and zinc.

53. *Stannic Chloride.*—A concentrated solution is used for establishing the presence of cæsium and for modifying several reactions.

54. *Platinic Chloride*, a slightly acidulated solution (chloroplatinic acid, $1:10$).—The best reagent for potassium and ammonium. It should not deposit octahedral crystals during evaporation.

55. *Potassium Chloroplatinate.*—A saturated solution. An excellent reagent for detecting and distinguishing rubidium and cæsium.

56. *Platinic Sulphate*, a solution of $1:20$.—It is prepared of sufficient purity by evaporating several times a solution of platinic chloride mixed with a slight excess of sulphuric acid, until a drop tested with sulphate or nitrate of potassium will deposit exclusively *prismatic* crystals (*no*

octahedra). A mixture of platinic sulphate with nitrate or sulphate of potassium affords a characteristic test for distinguishing chlorine, bromine, and iodine. Of other reagents may be mentioned *aluminium nitrate*, used with cæsium chloride in testing for sulphuric acid; *cadmium nitrate*, used in testing for sulphuretted hydrogen; *starch*, used in testing for iodine and bromine; *litmus*, in testing for acid and alkaline reaction; *congo red* and *malachite*

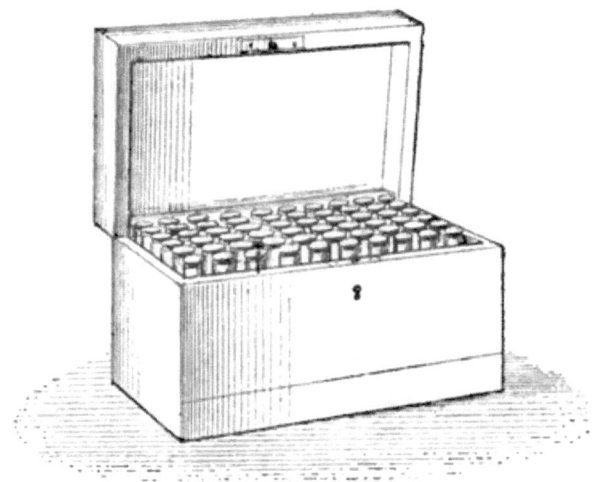

FIG. 3.—Box for reagents.

green, in staining tests. Some other reagents will be mentioned incidentally.

The reagents are kept in small tubes (50 mm. long 9 to 10 mm. wide), closed with ground stoppers or with India-rubber. These tubes are slid into holes, drilled in a block of light wood, which is fitted in a small box, provided with a drawer for accessories—platinum wires, spatula, spoons, forceps, capillary tubes, strips of iron, zinc, tin, filtering-paper, etc. Such a box, containing 60 tubes, can be made very small—height 8 cm., breadth 9, length 13 cm. Flat stoppers are convenient for lettering or numbering with a diamond, or for pasting on labels, which are finished by varnishing. The dimensions must in this case be made a little greater: 8 : 10 : 15 cm. For various reasons it is advisable to have another box of smaller dimensions for volatile reagents. The tubes in this box are somewhat greater, 60 : 15 mm. Along with these, the ebonite tubes for hydrofluoric acid and for ammonium fluoride can be placed. For travelling, the whole apparatus, microscope included, may be packed in a small valise.

V. REACTIONS

§ 1. Potassium

a. Precipitation with platinum chloride. Limit: 0.5 μgr. of K (1 μgr. = 0.001 milligr.) [1]

b. Precipitation with phosphomolybdic acid. Limit: 0.3 μgr. of K. [2]

c. Precipitation with sulphate of bismuth. Limit: 0.2 μgr. of K.

a. A solution of platinum tetrachloride (1 : 10) is employed as reagent. It should not deposit octahedral crystals during evaporation. A small drop is put into the liquid that has to be tested. The reaction of the latter must be neutral or slightly acid. From concentrated solutions the compound K_2PtCl_6 will come down immediately as a yellow crystalline powder; from dilute solutions beauti-

---

[1] Behrens, Mikroch. Method. p. 22; also Ann. de l'Éc. polyt. de Delft, i. 193.

[2] Behrens, l.c. p. 22; Ann. de l'Éc. pol. i. 193.

ful yellow octahedra are obtained, measuring from 10 to 70 μ. Sometimes they look like hexagonal plates, in

Fig. 4.—Potassium chloroplatinate, ×90.

consequence of combination with the cube and of abnormal growth of two opposite facets of the octahedron. An excess of acid is injurious; it is remedied by adding sodium acetate or magnesium acetate. A splendid reaction, provided that ammonium, rubidium, and cæsium are not present.

b. Phosphomolybdic acid forms large octahedral crystals of a light yellow colour, dissolving freely in water. A large drop of the reagent is added to a drop of the liquid that has to be tested, after acidulating with nitric acid. The crystals of potassium phosphomolybdate $(K_3PO_4(MoO_3)_{10} + 3H_2O)$ resemble those of potassium chloroplatinate as regards their size and colour. They are strongly refractive. Sometimes facets of the cube, the octahedron and the dodecahedron can be recognised; generally spheroidal grains are developed, accumulating along the border of the drop. This reaction

Fig. 5.—Potassium phosphomolybdate, ×120.

is valuable for solutions, containing much free acid. It is to be borne in mind that ammonium, rubidium, and cæsium are precipitated in the same way as potassium.

c. A little nitrate of bismuth is dissolved in sulphuric acid, to which a small quantity of dilute nitric acid may be added, if a slow action is desired. A small drop of this solution is brought in contact with a concentrated drop of the liquid that has to be tested. Very soon colourless hexagonal discs ($K_3Bi(SO_4)_3$) will show themselves spreading from the point of contact. They develop slowly into crystals of rhombohedral appearance, measuring from 30 to 60 μ. This reaction can be utilised, wherever it is desired to trace sodium and potassium at the same time. See § 2, d.

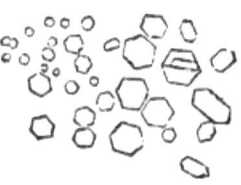

FIG. 6.—Double sulphate of potassium and bismuth, ×130.

To make the list complete, the test by means of *hydrofluosilicic acid*, pointed out by Boricky,[1] and the test by means of *cerous sulphate*, recommended by the author of this work, must be mentioned; neither of these, however, can rival those described above. With regard to accidental cubes of potassium silicofluoride, see § 39, a.

§ 2. Sodium

a. Precipitation with uranyl acetate. Limit: 0.8 μgr. of Na.[2]

b. Precipitation with uranyl acetate and acetate of magnesium. Limit: 0.4 μgr. of Na.[3]

c. Precipitation as silicofluoride of sodium. Limit: 0.16 μgr. of Na.[4]

d. Precipitation with sulphate of bismuth. Limit: 0.04 μgr. of Na.

a. Uranyl acetate is dissolved in dilute acetic acid.

---

[1] Boricky, l.c. p. 17.
[2] Streng, Ber. d. oberhess. Ges. f. Nat. u. Heilkunde, xxii. 258.
[3] Streng, l.c. xxiv.; Anleit. z. Bestimm. d. Mineral. p. 74.
[4] Boricky, l.c. p. 18.

The liquid that has to be tested for sodium must be evaporated, or at least strongly concentrated, because a dilution of 1 : 50 will render crystallisation doubtful. The crystals of the double acetate ($C_2H_3O_2Na . C_4H_6O_4UO_2$) are light yellow tetrahedra, neatly developed, measuring about 70 μ. Strong acids and ammonium salts are injurious to this beautiful reaction. According

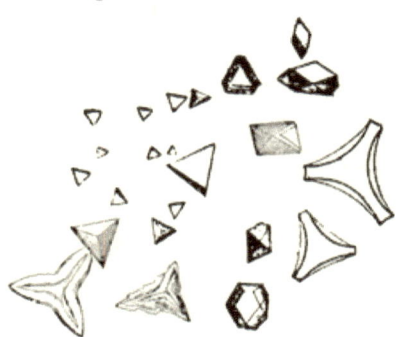

FIG. 7. — Double acetate of sodium and uranyl, at the right side crystals of the triple acetate of uranyl, sodium, and magnesium, ×50.

to Streng, it is hampered by platinum chloride; it will be prudent to bear this remark in mind when testing for potassium and sodium.

b. The reaction *a* is considerably modified by soluble compounds of magnesium, iron, cobalt, nickel, and copper. Triple acetates are produced, forming rhombohedral and scalenohedral crystals, which may simulate the octahedron, the tetrahedron, and the dodecahedron. The triple acetate of sodium, magnesium, and uranyl $C_2H_3O_2Na . C_4H_6O_4Mg . 3(C_4H_6O_4 . UO_2) + 9H_2O$ contains 1.48 per cent of Na. Its crystals are almost colourless, measuring about 120 μ. To the metals enumerated by Streng must be added beryllium, zinc, and cadmium.

c. Instead of hydrofluosilicic acid, recommended by Bořický, ammonium silicofluoride may be employed. It has the advantage of being easily purified and of giving no trouble in keeping. The reagent is added in a dry state to the moderately acidulated liquid. If much sodium is present, hexagonal rosettes (80 to 120 μ) are produced; from dilute solutions hexagonal plates and short prisms (70 μ), capped by a short pyramid, will separate. A feeble rose tint is observed in all crystals reaching 50 μ. This reaction is not injured by platinum chloride.

FIG. 8.—Silicofluoride of sodium, ×100.

d. According to Heintz,[1] bismuth sulphate will not form a crystalline compound with sodium sulphate. From concentrated solutions a precipitate of amorphous appearance is certain to be thrown down; but from dilute solutions, acidulated with nitric acid, the double salt is obtained in a crystallised state.

By addition of nitric acid the liquid can be caused to remain clear for some time. On heating it grows turbid by the precipitation of small rod-like crystals. Their length may attain 80 μ; it coincides with the direction in which they extinguish in polarised light. The transverse section of the crystals attains 16 μ; it is hexagonal, showing no double refraction. Bismuth sulphate may be utilised for tracing potassium and sodium in the same drop. For this purpose a little glycerine is added to the solution of the reagent; the mixture is spread in a thin layer on a clean slide. On touching it with a platinum wire, to which a few grains of the dry or pasty sample adhere, waiting about two minutes, and then heating to 50°, and allowing two minutes for cooling, crystallisation takes place. If the preparation should become too pasty, moisten it by the breath. Under a power of 150 the small rods of the sodium compound will be found accumulated around the spot touched with the wire; the discs of the potassium compound appear later and spread over a wider area.

[1] In Wurtz, Dictionn. de Chimie, Art. Bismuth.

Cerous sulphate[1] equals sulphate of bismuth as regards delicacy, the crystals of its compound with sodium sulphate, however, are dwarfed to such a degree that they cannot be distinguished with ease and certainty from those of the analogous potassium compound.

Potassium antimonate[2] will in some cases render good service. In dilute solutions the crystallisation of sodium antimonate (see § 49, *d*) is very slow; it can be accelerated by a small drop of alcohol. Calcium and magnesium must be absent.

§ 3. Lithium

a. Precipitation as fluoride. Limit: 0.25 μgr. of Li.
b. Precipitation as phosphate. Limit: 0.4 μgr. of Li.
c. Precipitation as carbonate. Limit: 0.36 μgr. of Li.[3]

a. The appearance of lithium fluoride (LiF) varies with the reagent employed for the precipitation. Potassium fluoride produces cubic crystals (15 to 20 μ), sodium fluoride produces cubes (20 μ), and close to the reagent hexagonal prisms (30 μ). Ammonium fluoride produces cubes (15 to 25 μ), or, if employed in excess, rectangular rosettes (80 to 100 μ). With sodium fluoride the limit is about 0.5 μgr. of Li; with an excess of ammonium fluoride the delicacy is doubled. Hydrofluosilicic acid (Bořický) is not to be recommended.

b. Precipitation with sodium phosphate necessitates heating to ebullition. Addition of sodium carbonate hastens the precipitation, but to make it complete boiling must be continued until all the liquid is evaporated. After washing with water the residue is found to consist of crystalline

---

[1] Behrens, Mikr. Meth. p. 24; Ann. de l'Éc. pol. i. 194.
[2] Haushofer, l.c. p. 99.
[3] Behrens, Mikr. M. p. 26; Ann. de l'Éc. pol. i. 196.

grains and imperfectly developed rhombs. A neutral solution, heated to incipient ebullition, after addition of sodium phosphate, yields rectangular crystals (Li_3PO_4), measuring from 10 to 20 μ. They exhibit strong double refraction, which is effaced in the direction of their edges.

c. The best precipitant is ammonium carbonate. From dilute solutions lithium carbonate (Li_2CO_3) will separate out along the margin of the drop, in the same way as calcium sulphate. From ammonium sulphate its needles are distinguished by adding a drop of water. Ammonium sulphate dissolves immediately, lithium carbonate remains a long time undissolved. With regard to other reagents it is to be noted, that solutions of lithium compounds are *not* precipitated by *platinum chloride; phosphomolybdic acid* and *bismuth sulphate* produce precipitates similar to those produced in solutions of potassium compounds, but less copious. *Potassium antimonate* provokes spherulitic crystallisations (30 to 50 μ), composed of slender needles. They are more soluble than the lenticular crystals (8 to 20 μ) of sodium antimonate.

Ammonium, see Nitrogen, § 53, *b*

§ 4. Cæsium

a. Precipitation as chlorostannate. Limit: 1.6 μgr. of Cs.[1]

b. Precipitation as silicomolybdate. Limit: 0.25 μgr. of Cs.

c. Precipitation as chloroplatinate. Limit: 0.1 μgr. of Cs.

a. Stannic chloride precipitates from moderately diluted

[1] Haushofer, l.c. p. 31 (from Mr. Sharples).

solutions well-defined colourless octahedra (30 to 40 μ) of the compound Cs_2SnCl_6. The precipitation is not sensibly modified by hydrochloric acid. Evaporate the sample, dissolve the residue in dilute hydrochloric acid and add stannic chloride. If this method is followed, there will be little chance of taking rubidium or potassium for cæsium, the chlorostannates of the former dissolving freely in hydrochloric acid. If the presence of ammonium is suspected, evaporation must be followed by calcination, ammonium chlorostannate being much less soluble than the chlorostannates of rubidium and potassium. The delicacy of the test may be increased by adding sodium iodide. The crystals of cæsium iodostannate have the form and size of the crystals of cæsium chlorostannate, but the yellow colour of potassium chloroplatinate. For more details see § 35, a.

FIG. 9.—Chlorostannate of cæsium, ×130.

b. According to Parmentier, cæsium and rubidium can be separated from potassium by means of silicomolybdic acid. A saturated solution of ammonium silicomolybdate [1] may be used as reagent. This solution is not precipitated by soluble compounds of potassium, sodium, and lithium. Ammonium chloride precipitates spheroidal crystals (8 to 20 μ), cæsium chloride produces yellow grains (2 to 6 μ), thallous salts yield a fine dust. All these precipitates bear a close resemblance to the analogous phosphomolybdates. Precipitation is promoted by free nitric acid.

[1] The reagent is prepared by mixing a solution of ammonium molybdate in nitric acid (reagent for phosphoric acid) with a solution of alkali silicate (water-glass) in dilute nitric acid. The mixture turns yellow, and on heating crystalline silicomolybdate separates out. The crystallisation is promoted by adding ammonium nitrate. The dense precipitate is rapidly washed and recrystallised from hot water.

c. Platinum tetrachloride, dissolved in 300 parts of water, may be very well employed for the same purpose as silicomolybdic acid. Evaporation must be excluded to prevent crystallisation of chloroplatinates of potassium and ammonium. Cæsium chloroplatinate (Cs_2PtCl_6) comes down in small neatly developed octahedra (3 to 5 μ), of the same colour as the analogous potassium compound (§ 1, *a*). Instead of a diluted solution of platinum tetrachloride, a saturated solution of potassium chloroplatinate may be used. See Rubidium, § 5, *b*.

§ 5. Rubidium

a. Precipitation as silicomolybdate. Limit: 0.7 μgr. of Rb.
b. Precipitation as chloroplatinate. Limit: 0.5 μgr. of Rb.

a. The crystals of rubidium silicomolybdate are far greater than those of the cæsium compound. They measure from 10 to 20 μ. Instantaneous precipitation is limited for rubidium chloride at a hundredfold dilution, for cæsium at a dilution of 1 : 300.

b. From a theoretical point of view stannic chloride and ammonium silicomolybdate will be considered as the best reagents for distinguishing cæsium and rubidium in the presence of potassium. In practice a saturated solution of potassium chloroplatinate is found far superior. By this reagent *thallium* is precipitated first of all. The precipitate is composed of very small octahedra (1 to 2.5 μ). Then follows *cæsium* chloroplatinate (2 to 6 μ), later *rubidium* chloroplatinate (8 to 20 μ), spreading over thrice the distance.[1] If evaporation is not prevented, crystallisation of

---

[1] Two or three small drops of the sample are evaporated on the same spot, a drop of the reagent is put upon the residue, spreading beyond it.

potassium chloroplatinate may set in at the end, beginning at the circumference of the drop and yielding crystals of 30 to 70 μ.

Precipitation of rubidium as tartrate, by means of tartaric acid or by means of acid sodium tartrate, is open to the objection, that much acid potassium tartrate may come down with the rubidium compound.

§ 6. Thallium

a. Precipitation with hydrochloric acid. Limit: 0.16 μgr. of Tl.[1]

b. Precipitation with potassium iodide. Limit: 0.03 μgr. of Tl.[2]

c. Precipitation as chloroplatinate. Limit: 0.008 μgr. of Tl.

a. Hydrochloric acid and soluble chlorides produce in solutions of thallous salts well-defined colourless cubes of the compound TlCl (10 to 15 μ). From concentrated solutions rectangular rosettes are thrown down, measuring from 50 to 100 μ. They are white by reflected light, nearly black by transmitted light. Thallous chloride dissolves at ordinary temperature in 400 parts of water;

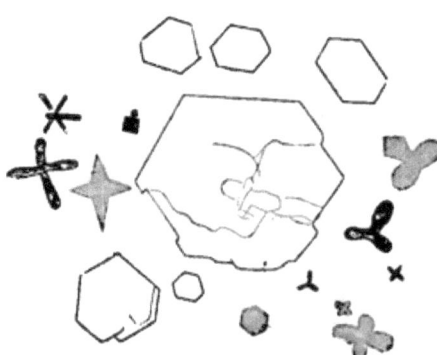

FIG. 10.—Thallous chloride and thallic chloride (the great clear scales), × 200.

---

[1] Behrens, Mikr. Meth. p. 34 ; Haushofer, l.c. p. 125.
[2] Behrens, l.c. p. 36 ; Haushofer, l.c. p. 125.

the solubility is considerably increased by heat. From a solution in hot water it crystallises in clear, sharply-cut cubes, characterised by very high refraction. It is far more soluble in sulphuric and nitric acids than in water. Sodium acetate may be employed to neutralise the injurious influence of these acids. The test with hydrochloric acid is not very delicate, but simple and characteristic.

b. Thallous iodide is much less soluble (1:4500) than the chloride. At the same time its crystals are much smaller, the largest rosettes attain no more than 20 μ. Their form is cubic, their colour a dark yellow, nearly opaque.

c. Platinum tetrachloride precipitates thallium from solutions in which potassium iodide fails to indicate the metal. The crystals are somewhat greater than those of the iodide, about half the size of the octahedra of cæsium chloroplatinate.

Phosphomolybdic acid produces similar crystals in strongly acidulated solutions. This test is, however, far inferior to that described in *c* in delicacy; this inferiority is sometimes counterbalanced by the advantage of being able to control the result by adding an excess of alkali. Thallous phosphomolybdate is dissolved by caustic potash or soda; from the solution *thallous molybdate* (Tl_2MoO_4) will speedily separate out. For the characteristics of this compound see § 57, *b.*

The reactions, proposed by Haushofer, founded upon the use of oxalic acid and of potassium bichromate,[1] are likewise of inferior delicacy. The description of *thallous chromate* in the work of Haushofer is not complete. A yellow powder, it is true, is thrown down, but in the course of a few minutes it is changed to beautiful needles of the bichromate $Tl_2Cr_2O_7$. This reaction might certainly be

[1] Haushofer, l.c. p. 127.

utilised, but for the superior delicacy of the chloride and iodide tests.

Of *thallic* compounds the chloride $TlCl_3$ may be mentioned, as it sometimes occurs among the products of microchemical reactions. It is more soluble than thallous chloride; from its solution in hot water it crystallises in colourless or grayish hexagonal plates, measuring from 20 to 50 μ. Thallic chloride is produced wherever a thallous compound is brought into connection with free chlorine. It should be borne in mind that some thallous compounds, among others the molybdate and the tungstate, resemble the thin flakes of thallic chloride.

§ 7. Silver

a. Precipitation as chloride.   Limit: 0.1 μgr. of Ag.[1]
b. Precipitation as chromate.   Limit: 0.15 μgr. of Ag.[2]

a. Amorphous chloride of silver is useless for microchemical tests. It may be made to crystallise by heating it with hydrochloric acid (Streng), or by dissolving it at ordinary temperature in caustic ammonia (Haushofer). From the solution in hydrochloric acid octahedral crystals are obtained on cooling; from the ammoniacal solution, cubes and combinations of the cube with facets of the octahedron will separate out, the crystallisation keeping pace with the volatilisation of the ammonia. I prefer working with ammonia. By its aid

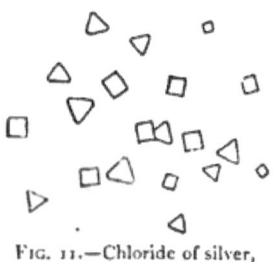

Fig. 11.—Chloride of silver, ×200.

---

[1] Streng, Ber. d. oberhess. Ges. xxiv. 54.
[2] Haushofer, l.c. p. 118.

silver chloride can be easily separated from many other insoluble compounds, and it is less troublesome under the microscope than strong hydrochloric acid. High powers must be employed to distinguish the little crystals (8 to 20 μ); for very small quantities of silver a power of 300 must be employed. This inconvenience is counterbalanced by several good qualities. The test is of great value for tracing silver among a variety of other metals. It is not impaired by compounds of alkali metals; even chromates, phosphates, and arseniates do not interfere with it in any way. Chloride of zinc and antimonous chloride have a tendency to increase the size of the crystals, and to change their form, producing six-sided plates of 20 to 50 μ. Of platinum tetrachloride only very little is taken up; it occasions the growth of complicated cruciform rosettes, measuring from 60 to 90 μ. Mercuric chloride and ammonium molybdate have an injurious effect on this test. Stannic chloride spoils it entirely.

b. According to Haushofer, potassium bichromate precipitates thin brownish and blackish rods from neutral solutions of silver. Numerous trials with slightly acidulated solutions of varied concentration, precipitated with bichromate of potassium, and at other times with bichromate of ammonium, have invariably resulted in the production of large prisms (triclinic system) of a brilliant brownish-red. In a few minutes parallelograms of 30 to 70 μ, and needles of 300 μ can be obtained; the needles will grow in a quarter

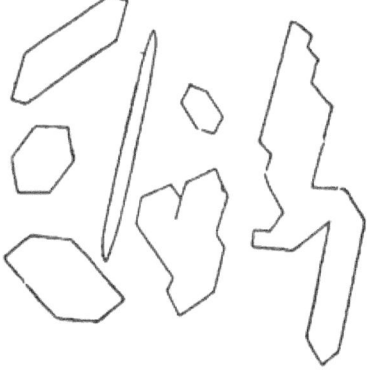

FIG. 12.—Bichromate of silver, ×60.

of an hour to 2000 μ. Recrystallisation may be effected by heating with very dilute nitric acid. Instead of the seemingly monoclinic bichromate $Ag_2Cr_2O_7$, the orthorhombic chromate Ag_2CrO_4 is formed if a great quantity of sulphate is present. See § 44, *a*.

This reaction is superior to that described in *a*, provided that soluble chlorides are absent, and that other metals precipitated by bichromate are not present in any considerable quantity.

Other compounds of silver have been proposed for microchemical testing: the *carbonate*, the *arsenite*, the *phosphate*, the *tartrate*, and the *acetate* by Haushofer,[1] the *arsenate* by Streng.[2] All these compounds dissolve freely in ammonia and in strong acids; the arsenate, which yields reddish crystallites with three spokes, is not dissolved by acetic acid. The acetate crystallises readily, yielding large pearly plates. It will certainly be utilised in microchemical analysis, if a test for acetic acid is wanted.

§ 8. Magnesium

a. Precipitation as ammonium-magnesium phosphate. Limit: 0.0012 μgr. of Mg.[3]

b. Precipitation as magnesium pyroantimoniate.

a. As reagents sodium phosphate or microcosmic salt can be used. It is a good plan to heat after adding ammonium chloride and an excess of ammonia. A grain of sodium phosphate is put into the warm liquid. Heating has been recommended by Streng; it greatly assists in producing well-developed crystals. From solutions con-

---

[1] Haushofer, l.c. p. 120. [2] Streng, Anl. z. Best. d. Min. p. 85.
[3] Behrens, Mikr. Meth. p. 21.

taining more than 0.5 per cent of magnesium, rudimentary crystals are chiefly precipitated resembling the letter X. From very dilute solutions hemimorphic crystals (10 to 20 μ) of the orthorhombic system separate out. Besides the compound $NH_4MgPO_4 + 6H_2O$, there are isomorphous double phosphates of the same type, containing Fe, Mn, Co, Ni in the place of Mg. Iron and manganese are thrown down as hydroxides by waiting about a minute after supersaturation with ammonia before adding sodium phosphate. Cobalt may be precipitated with iron and manganese by adding hydrogen peroxide to the ammoniacal solution. (See § 11 c; § 12 b.) It is to be noted that the test for magnesium sometimes fails for want of ammonium chloride, or of caustic ammonia.

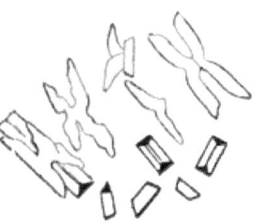

FIG. 13.—Ammonium-magnesium phosphate, ×200.

b. Magnesium pyroantimoniate ($MgH_2Sb_2O_7 + 9H_2O$) is precipitated by potassium antimoniate in glittering six-sided prisms. Sodium and calcium are precipitated at the same time.

Potassium ferrocyanide does not precipitate magnesium from neutral or acid solutions. If ammonia is added, slender needles and thin plates of a double salt are thrown down. The *tartrate* exhibits an analogous behaviour. The delicacy of these tests is one-tenth of that found in *a*.

Oxalic acid will not produce any precipitate in neutral solutions. If a great quantity of acetic acid is added, slender needles of magnesium oxalate separate out. Solutions containing magnesium and zinc will sometimes yield hexagonal plates of a double oxalate.

Potassium oxalate produces a granular crystallisation in solutions of moderate concentration. It is dissolved by an excess of the reagent, unless the solution be highly concentrated.

Sodium bicarbonate will leave solutions of magnesium salts clear for a long time. During evaporation, light flakes will separate out. If calcium be present, a granular precipitate is formed instead of the sharply-defined rhombohedra of calcium carbonate.

Fluosilicate of magnesium, recommended by Bořický,[1] exhibits rhombohedral crystals. It is highly soluble, like calcium fluosilicate.

§ 9. Beryllium

Precipitation with potassium oxalate. Limit: 0.08 μgr. of Be. Limit of instantaneous precipitation, 1 part of beryllium sulphate in 600 parts of water.

A slight addition of potassium oxalate will precipitate X-shaped crystals of beryllium oxalate, soluble in about 100 parts of water. An excess of the reagent will produce stout highly refractive monoclinic prisms of potassium-beryllium oxalate ($BeC_2O_4 \cdot K_2C_2O_4$), often twinned in the same manner as gypsum. A great

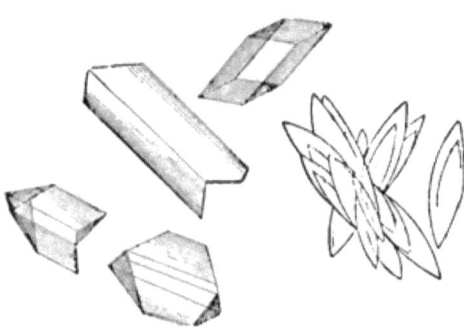

FIG. 14.—Potassium-beryllium oxalate, × 120.

excess of the reagent leads to sheaf-like combinations of pale rhombs and leaflets. Ammonium oxalate is less fitted for this test; sodium oxalate will not produce any

[1] Bořický, l.c. p. 21.

crystallisation. Potassium-beryllium oxalate dissolves freely in hot water. The solution may remain a long time in a state of supersaturation. Crystallisation may be started by crushing a grain of potassium oxalate in the drop. By some compounds the aspect of the crystals of potassium-beryllium oxalate is modified. If mercuric chloride is present, long prisms are chiefly formed (120 to 200 μ), exhibiting vivid colours between crossed nicols. Angle of extinction $= 40°$. If zinc or magnesium be present, the size of the crystals is reduced. A small quantity of zinc and a considerable quantity of magnesium are combined with the beryllium in a compound oxalate. For testing with potassium oxalate, solutions of beryllium must be neutral or slightly acidulated with acetic acid. The test is not sensibly injured by the presence of aluminium or iron. If ammonium carbonate has been employed for separating beryllium from other metals, ammonium salts must be driven off by heat before testing for beryllium.

The tests proposed by Haushofer [1]—crystallisation of beryllium chloroplatinate and of beryllium sulphate—are of little value. For these tests beryllium must be separated from all other metals, and when this condition is fulfilled their want of delicacy is found a serious inconvenience. Beryllium sulphate dissolves freely in water; beryllium chloroplatinate is deliquescent.

With greater success the second of the reactions pointed out by Streng for sodium (§ 2, b) can be applied to beryllium. Acetate of uranyl will form with acetate of beryllium and a very small quantity of sodium acetate pale yellow rhombohedra, growing to 200 μ. Rapid growth will produce skeletons of tetrahedral aspect. If too much sodium acetate is added the ternary compound is suppressed

[1] Haushofer, l.c. pp. 23, 24.

by rapid growth of bright yellow tetrahedra of sodium-uranyl acetate. In this case precipitate with caustic ammonia, absorb the liquid with a strip of filtering-paper, and redissolve in acetic acid.

The method of Rössler, viz. precipitation of ammonium-beryllium phosphate by boiling a neutralised solution of this compound in hydrochloric acid, is not fitted for microchemical use. The precipitate is dense, but not at all crystalline.

§ 10. Manganese

a. Precipitation with oxalic acid. Limit : 1 μgr. of Mn. If potassium oxalate is employed, the limit is carried to 0.5 μgr. of Mn.[1]

b. Precipitation as ammonium-manganous phosphate. Limit : 0.3 μgr. of Mn.

c. Precipitation as peroxide. Limit : 0.2 μgr. of Mn.

a. From solutions containing a small quantity of free acetic acid characteristic stars (100 to 120 μ) of the compound $MnC_2O_4 + 3H_2O$ are precipitated by oxalic acid. These stars are formed by three rods crossing each other. Each rod is composed of two twinned slender prisms, with basal angles of 60°. By great quantities of zinc, nickel, or cobalt the development of these stars is interfered with. Strong acids prevent the precipitation of manganese by oxalic acid; great quantities of alkali salts likewise prove injurious. In the last case addition of an excess of ammonia will be found useful. It produces stars and fringed rods ($MnC_2O_4 + (MnN_2H_6)C_2O_4 + 6H_2O$) of smaller size (50 μ), under circumstances where the test with oxalic acid has failed. After some minutes the out-

---

[1] Haushofer, l.c. p. 96.

lines of the crystals become faint. They are, however, restored by addition of a small drop of ammonia.

b. It is essential for the test that oxidation of the ammoniacal solution be precluded. The slide with the drop of acid manganous solution is slightly heated, and a mixture of ammonia and sodium phosphate is added. The hemimorphic crystals attain a length of 40 μ, otherwise they resemble the analogous compound of magnesium.

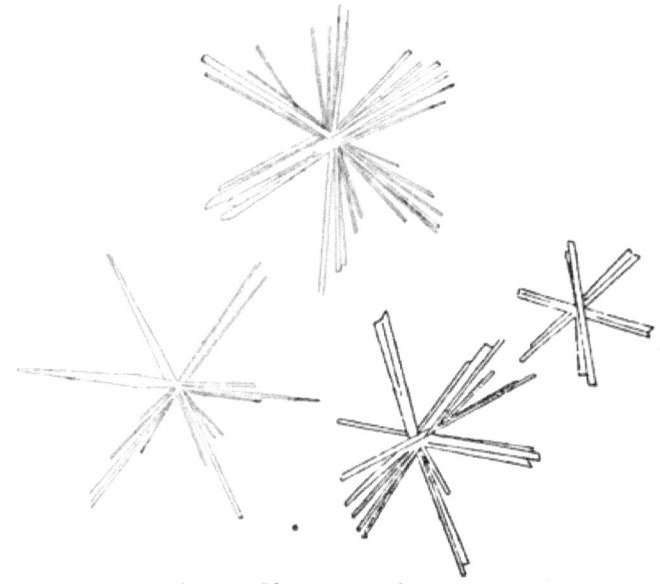

FIG. 15.—Manganous oxalate, × 120.

By their adhesion to the glass, washing is made easy. In washed crystals the presence of manganese is established by adding hydrogen peroxide and caustic alkali. The crystals are stained dark brown without undergoing deformation.

c. From magnesium, zinc and nickel, manganese and cobalt can be separated by adding hydrogen peroxide to a solution in caustic ammonia. From cobalt, manganese is separated by boiling with nitric acid and potassium chlorate (Hampe's method). After evaporation of the acid a film of dark brown peroxide is left, which will bear cautious washing and afford excellent material for testing. Generally the dry way will be chosen—testing on platinum with sodium carbonate. It is far superior to the wet way as regards delicacy, and highly characteristic.

§ 11. Cobalt

a. Precipitation with potassium nitrite. Limit: 0.1 μgr. of Co.[1]

b. Precipitation with ammonium-mercuric thiocyanate. Limit: 0.3 μgr. of Co.

c. Precipitation as ammonium-cobaltous phosphate. Limit: 0.02 μgr.

d. Precipitation as chloride of purpureocobalt. Limit: 0.2 μgr. of Co.

a. This reaction can be employed in acid and in ammoniacal solutions. In the latter an intense brown colour is produced. If afterwards acetic acid is added, yellow grains (2 to 4 μ) of the compound $Co_2(NO_2)_6 \cdot 6KNO_2 + 3H_2O$ are thrown down, looking almost black in transmitted light. In hot solutions these grains may attain 20 μ, when they will appear as dark yellow cubes and octahedra. This reaction is very useful for separating cobalt from nickel, the more so as the precipitate settles firmly on the glass. The delicacy

FIG. 16.—Double nitrite of potassium and cobalt, × 300.

[1] Streng, Ber. d. oberh. Ges. xxiv. 58.

can be doubled by adding cæsium chloride; by thallium nitrate the limit is carried beyond 0.02 µgr.

b. A solution of mercuric thiocyanate in ammonium thiocyanate is a convenient and characteristic reagent for cobalt. If no more ammonium thiocyanate is employed than is required for dissolving the mercuric thiocyanate, the reagent will precipitate insignificant needles of the compound $2(Hg(CyS)_2) \cdot Co(CyS)_2$. Half as much more of the solvent must be added to obtain the full effect. The crystals of the double thiocyanate $Hg(CyS)_2 \cdot Co(CyS)_2$ are orthorhombic. Isolated crystals are, however, seldom seen; they are generally grouped in irregular fashion. Their colour is a magnificent blue, of such intensity as to approach black in crystals of moderate thickness. Strong acids impair the reaction, and in ammoniacal solutions it will not succeed at all. Ammonia destroys the colour of the crystals; by acetic acid it is restored. An excess of acetic acid has no sensible influence on the reaction.

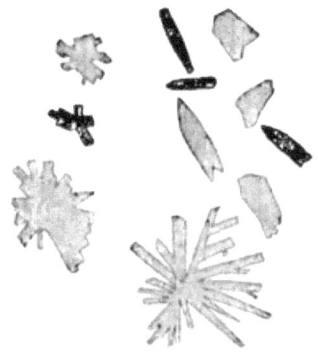

Fig. 17.—Mercuric cobaltous thiocyanate, ×60.

With caustic soda and hydrogen peroxide the mother liquor grows brown, but the crystals are scarcely attacked. Ammonium-mercuric thiocyanate has no perceptible action on solutions containing nickel, manganese, or magnesium. In solutions of zinc or cadmium it produces colourless rhombs. If cobalt and zinc or cobalt and cadmium are present, compound *pale blue* crystals are produced. In cupric solutions rhombs and needles of a greenish-yellow are produced. In mixed solutions of cobalt and copper no compound crystals are formed. This combination can, however, be brought about by the addition of a great quantity of zinc or cadmium. It is characterised by a particular tint, a brownish-violet. These variations can be turned to profit for detecting several metals at once. Addition of zinc acetate also proves helpful if cobalt has to be traced in solutions containing a great quantity of alkali salts.

c. The operation is carried out in the same way as for manganese (§ 10, *b*). The hemimorphic crystals are slender. They may attain a length of 100 μ if their growth is favoured by a great excess of ammonia and ammonium chloride. By caustic alkali and hydrogen peroxide they are stained dark brown. This reaction may be employed to distinguish cobalt from nickel. Another expedient for establishing the presence of cobalt in the double phosphate is furnished by application of the preceding reaction (§ 11, *b*). Wash with a drop of water, add acetic acid, and test with ammonium-mercuric thiocyanate. Addition of zinc acetate will be found useful.

d. The solution is treated as in *c*, except that in the place of sodium phosphate a small quantity of potassium permanganate is added. Sometimes, at this stage, stout brownish crystals will separate out, probably chloride of luteocobalt. They are decomposed by the final heating with a slight excess of hydrochloric acid, which precipitates small violet prisms and octahedral crystals of the chloride of purpureocobalt ($Co_2Cl_6 . 10NH_3$). It can be recrystallised from very weak hot hydrochloric acid. In acid of moderate concentration it is almost insoluble. Magnesium, zinc, and nickel may be separated from cobalt and manganese by converting the latter into peroxides. This can be done by adding hydrogen peroxide to a solution in caustic ammonia, or by heating the nitrates till the mass is blackened. Dilute nitric acid will leave the peroxides of cobalt and manganese undissolved. With regard to the precipitation with oxalic acid, see § 12, c.

§ 12. Nickel

a. Precipitation with potassium nitrite and acetate of lead. Limit: 0.008 μgr. of Ni.

b. Precipitation as ammonium-nickelous phosphate. Limit: 0.01 μgr. of Ni.

c. Precipitation with oxalic acid. Limit: 1 μgr. of Ni.[1]

a. Potassium nitrite will not produce any visible effect in solutions of nickel. If potassium nitrite and acetate of lead are added to an ammoniacal solution, a heavy white precipitate of insoluble compounds of lead is formed. By a small drop of acetic acid this precipitate is changed to a yellow powder of the compound $K_2PbNi(NO_2)_6$. A slight heating will favour the reaction. This compound can be recrystallised from hot diluted acetic acid. In this respect it resembles the potassium cobalt nitrite, both in respect to form and colour. Soluble compounds of barium, strontium, and calcium may be employed instead of those of lead, but their action is slower. If the sample contains abundance of soluble sulphates, calcium acetate must be employed, although it makes the test more tedious, and the precipitate less fitted for washing.

FIG. 18.—Triple nitrite of potassium, nickel, and lead, ×200.

b. A good deal of ammonia and ammonium chloride

[1] Haushofer, l.c. p. 64.

must be employed. The test will succeed better with ammonium chloride and sodium bicarbonate. The double phosphate of nickel forms shorter crystals than the analogous compound of cobalt; they have often the appearance of square plates. By caustic alkali and hydrogen peroxide they are not stained. The same holds true of ammoniacal solutions mixed with hydrogen peroxide.

c. Oxalic acid has a very slow action on solutions of nickel. Even after long waiting the precipitation is found incomplete. The delicacy is only 0.4 of that obtained with solutions of cobalt. Oxalate of nickel is thrown down as fine dust, like that of copper, clouding the whole drop. An excess of potassium oxalate will produce pyramidal crystals. Ammoniacal solutions of cobaltous oxalate can be kept clear for several hours by adding a strong dose of ammonium chloride, while oxalate of nickel will separate out under the same circumstance, not having undergone any appreciable change. Laugier long ago based upon this different behaviour of the two oxalates a method for separating cobalt from nickel. It must be abandoned for reason of its tediousness, although it has been again recommended by Haushofer.

§ 13. Zinc

a. Precipitation with sodium bicarbonate. Limit: 0.01 μgr. of Zn.

b. Precipitation with oxalic acid. Limit: 0.1 μgr. of Zn.[1]

c. Precipitation with ammonium-mercuric thiocyanate. Limit: 0.1 μgr. of Zn.

d. Precipitation with potassium ferricyanide. Limit: 0.05 μgr. of Zn.

[1] Haushofer, l.c. p. 151.

a. From acid solutions containing zinc and cadmium, sodium bicarbonate will immediately precipitate round highly refracting grains (2 μ) of cadmium carbonate. Light flakes of zinc carbonate are at once perceived, slowly changing to small heaps of colourless, highly refractive tetrahedra (5 to 15 μ) of the compound $8ZnCO_3 \cdot 3Na_2CO_3 + 8H_2O$ if the reagent is added in excess. The characteristic

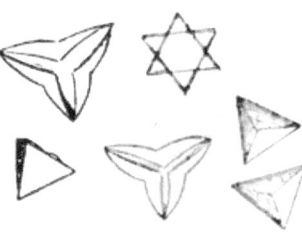

FIG. 19.—Double carbonate of zinc and sodium, ×300.

form of its crystals, and their tendency to cling to the glass, make the double carbonate of zinc and sodium of great value for tracing zinc and for separating it from other metals. Potassium carbonate yields larger tetrahedra, with a tendency to twinning. Lithium carbonate exhibits these peculiarities in more marked degree. Ammonium salts are highly injurious; small globules (6 to 10 μ) are very slowly produced. Magnesium salts spoil the reaction; instead of well-defined crystals a fine granular precipitate is produced. Cadmium is likewise injurious if its proportion rises above 5 per cent.

Ammoniacal solutions of zinc show a different behaviour. Cadmium carbonate is also precipitated immediately after addition of sodium bicarbonate. The tetrahedra of the zinc compound will appear about a minute later. Near the border of the drop strongly refractive grains are perceived, developing rapidly into tetrahedra (10 to 25 μ). Under these circumstances cadmium, cobalt, nickel, and calcium are deprived of their injurious influence. If too much powdery precipitate is formed, draw a portion of the clear ammoniacal solution aside with a platinum wire, cut the narrow channel with a small roll of filtering-paper, and let the crystallisation go on in the clear drop. Magnesium has the same injurious influence as in acid solutions; a similar influence is exercised by copper. A small quantity of magnesium may be precipitated by cautiously adding sodium phosphate (§ 8, *a*). If magnesium is present in great quantity it will be more safe and expeditious to precipitate zinc and cadmium as metals by means of magnesium. After rapid washing, the metallic sponge is treated with hot acetic acid, which chiefly dissolves zinc, and afterwards with hydrochloric acid, to dissolve cadmium. In examining alloys the following modification of the test will be found useful. The metals are converted into oxides by calcining their nitrates. *Zinc* oxide is dissolved by evaporating with a solution of caustic soda and by washing the alkaline mass with water. The precipitation of sodium zinc carbonate is effected by adding ammonium carbonate. If the quantity of zinc is small, the solution must be left to dry up. The tetrahedra are brought to light by dissolving the alkali salts in a drop of water.

b. Oxalic acid will precipitate small prismatic crystals (20 to 25 μ) of the compound $ZnC_2O_4 + 2H_2O$, even from solutions containing a considerable quantity of free hydrochloric acid. The crystals are a little longer than those of strontium oxalate; they are seldom found with sharp edges and angles. Soluble oxalates act more rapidly than oxalic acid, and will produce smaller crystals. With an excess of potassium oxalate pyramidal crystals can be obtained, like those of potassium-cobaltous oxalate, far more soluble than pure zinc oxalate. Zinc oxalate dissolves freely in caustic ammonia; as the solvent is volatilised, beautiful curled rosettes (200 to 700 μ) of an ammoniacal compound will be formed. From solutions containing much magnesium, oxalic acid sometimes precipitates pale hexagonal plates

(40 to 60 μ) of a double oxalate. Cadmium has no influence on the form of zinc oxalate; it is indeed entirely masked. On the other hand, cadmium and magnesium destroy the crystallisation of zinc oxalate from ammoniacal solutions.

c. The primary form of zinc-mercuric thiocyanate is a rectangular rod (orthorhombic system). It is sometimes seen, if very dilute solutions are precipitated, with a great excess of ammonium thiocyanate. Forked or dendritic aggregates will be generally formed, often curved and pinnated. This crystallisation is not injured by magnesium. By cadmium it is partially masked; it tends to make the form of the zinc compound less complicated. For compound thiocyanates containing mercury, zinc, and cobalt, see § 11, *b*; for mercury, zinc, and copper, see § 23, *c*.

d. Potassium ferrocyanide produces a colourless jelly, potassium ferricyanide will precipitate a yellow powder ($Zn_3Fe_2Cy_{12}$), or, from very dilute solutions, small yellow cubes (5 to 12 μ). Cadmium exhibits the same behaviour. Magnesium has no influence on this test.

e. Zinc phosphate dissolves freely in ammonia, without combining into a double phosphate, totally differing in this respect from the *arsenate*, which combines with ammonia, yielding hemimorphic crystals, resembling in form and dimensions those of ammonium magnesium phosphate. They will sometimes occur in the examination of arsenical ores; they may also be utilised in testing for arsenic acid, whenever the use of calcium compounds (§ 51, *a*) is excluded. Cadmium exhibits the same behaviour.

The *chromate* of zinc resembles strontium chromate in all respects, excepting its solubility in caustic ammonia, which it has in common with cadmium chromate. This reaction is not influenced by magnesium.

§ 14. Cadmium

a. Precipitation with sodium bicarbonate. Limit: 0.01 μgr. of Cd.

b. Precipitation with oxalic acid. Limit: 0.34 μgr. of Cd.[1]

c. Precipitation with ammonium-mercuric thiocyanate. Limit: 1 μgr. of Cd.

d. Precipitation with potassium ferricyanide. Limit: 0.03 μgr. of Cd.

a. Precipitation with sodium bicarbonate, as described above (§ 13, *a*), is not conclusive, unless it be carried out in an ammoniacal solution, treated with sodium phosphate to eliminate magnesium, manganese, etc. If the quantity of cadmium carbonate is not too small, its presence may be established by the following reaction (§ 14, *b*).

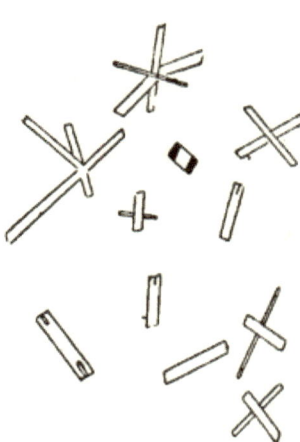

FIG. 20.—Oxalate of cadmium, × 120.

b. Oxalic acid precipitates the compound $CdC_2O_4 + 3H_2O$. If no zinc is present, the crystals of cadmium oxalate are seen as long parallelograms and rectangles of the monoclinic system (40 to 80 μ), characterised by oblique extinction (24°). The crystallisation is injured by 10 per cent of zinc; with 30 per cent of zinc the cadmium is masked. Evaporation of a solution of cadmium oxalate in ammonia yields large rods and plates of an ammoniacal compound; if zinc is present, only spherulitic grains are deposited. By free acids

[1] Haushofer, l.c. p. 53.

the crystallisation of cadmium oxalate is retarded; likewise by soluble compounds of aluminium, chromium, and iron. At the same time the length of the crystals is increased to 120 μ, and crossed twins will appear in great numbers. By means of aluminium sulphate and acetic acid, cadmium can be kept dissolved, while the main part of the zinc is precipitated. The precipitation of cadmium may be effected later by adding potassium oxalate. It is, however, more satisfactory to combine the reactions *a* and *b* in the following way:—*Cadmium* is precipitated from ammoniacal solution with sodium bicarbonate, the mother liquor containing the main part of the *zinc* is run off, and finally a small drop of water and a grain of oxalic acid are added. Characteristic crystals of cadmium oxalate will very soon be produced. Separation by caustic potash is less satisfactory; the same may be said of precipitation by metallic magnesium and extraction of the zinc with acetic acid.

c. Ammonium-mercuric thiocyanate will precipitate large rectangular prisms. By zinc, cadmium is masked. The crystals are stained light blue if a trace of cobalt is present.

d. Cadmium ferricyanide ($Cd_3Fe_2Cy_{12}$) forms yellow cubes (3 to 4 μ) perfectly resembling the crystals of the ferricyanide of zinc.

With regard to the *chromate*[1] it is to be noted that it resembles the chromates of zinc and of strontium. Metallic cadmium may be precipitated from acid solutions by metallic zinc.[2]

Metallic magnesium will precipitate rapidly both cadmium and zinc. Zinc may be extracted from the spongy metal by heating with acetic acid, the separation is, however, not complete.

[1] Haushofer, l.c. p. 54.   [2] Ibid. p. 52.

§ 15. Cerium

a. Precipitation with sodium sulphate. Limit : 0.02 μgr. of Ce.[1]

b. Precipitation with sodium carbonate. Limit: 0.05 μgr. of Ce.

c. Precipitation with oxalic acid. Limit : 0.04 μgr. of Ce.[2]

d. Precipitation with potassium ferrocyanide. Limit : 0.1 μgr. of Ce.

a. This test is of great delicacy and is very valuable for separations, although the action of sodium sulphate is rather slow; it may be accelerated by a moderate heat. The dense white precipitate ($3CeSO_4 \cdot Na_2SO_4 + 2H_2O$) is composed of small lenticular crystals (5 μ). A slight excess of sulphuric acid is favourable; any considerable excess of strong acids has an injurious influence, which ought to be neutralised by addition of magnesium acetate. Potassium sulphate precipitates small discs (4 to 6 μ). Thallous sulphate acts more slowly; it produces six-sided plates (orthorhombic) of considerable size (70 to 100 μ). All these double sulphates recur, with insignificant modifications, in the series of the compounds of lanthanum and didymium.

Fig. 21.—Double sulphate of sodium and cerium, ×500.

b. Sodium carbonate produces a flaky precipitate ($CeCO_3 + 3H_2O$) in solutions of cerous compounds. With an excess of the reagent the appearance of the precipitate is after some time changed. It becomes dense and crystalline, prickly knobs are formed (20 to 30 μ), and at the circumference of the drop pointed rhombs (20 to 30 μ), characteristic of cerium. The growth of these rhombic

---

[1] Behrens, Mikr. Meth. p. 24.   [2] Haushofer, l.c. p. 42.

leaflets is not impaired by the presence of lanthanum and didymium. The oxalate and the double sulphates of cerium are likewise decomposed by sodium carbonate more slowly, however, than the nitrate.

c. Oxalic acid produces a powdery precipitate ($CeC_2O_4 + 3H_2O$) in neutral solutions of cerous salts. This precipitate changes speedily to slender needles, to prickly spherulites, and to little crosses and squares (50 to 120 μ).

d. Potassium ferrocyanide precipitates transparent grains (4 to 6 μ), round, or square with rounded angles. With sodium-cerous sulphate the action is slower.

Cerous *formate*, recommended by Haushofer, yields large crystals resembling dodecahedra, but it is difficult to crystallise. It dissolves readily in hot water, and is apt to give supersaturated solutions. For tracing cerium it is not suitable, but it is valuable for separating it from calcium and from some other bivalent metals. These remarks apply also to the formates of lanthanum and didymium.

Calcination of cerium oxalate at a low red heat yields a rust-coloured powder of ceric oxide, not soluble in dilute nitric acid. The same compound may be prepared from the double sulphate of cerium and sodium by calcination with sodium carbonate and by a careful washing of the calcined mass. By heating with sulphuric acid ceric oxide is converted into cerous sulphate. These remarks indicate a method for an approximate separation of cerium from lanthanum and didymium. Diluted nitric acid will dissolve the oxides of the latter, while reddish ceric oxide is left.

§ 16. Lanthanum

a. Precipitation with sodium sulphate. Limit: 0.04 μgr. of La.

b. Precipitation with sodium carbonate. Limit: 0.06 μgr. of La.
   c. Precipitation with oxalic acid. Limit: 0.06 μgr. of La.
   d. Precipitation with potassium ferrocyanide. Limit: 0.1 μgr. of La.

   a. In dilute solutions the precipitation is slow at ordinary temperature. Small rods (10 to 12 μ) with rounded ends are formed around the drop. The liquid is very sensitive to heating; very soon it grows milky and deposits a permanent powdery precipitate $Na_2La(SO_4)_2$. With thallous sulphate six-sided plates are produced, perfectly resembling those described in § 15, *a.*

   b. Sodium carbonate yields a somewhat gelatinous precipitate, which begins to change after five minutes, if a considerable excess of sodium carbonate is employed. After half an hour all is transformed into prickly knobs (40 to 70 μ), composed of pale plates, seldom found isolated or twinned ($LaCO_3 . 2LaO + H_2O$).

   c. Oxalic acid precipitates very small grains (3 μ) of neutral oxalate (LaC_2O_4), speedily changing to needles and pointed rhombs (60 μ) of an acid oxalate. The needles are generally grouped in sheafs (40 to 100 μ).

   d. A minute addition of potassium ferrocyanide yields small rhombs (50 μ), a larger amount of the reagent will produce a double compound, appearing in colourless hexagonal plates (30 to 40 μ), exhibiting no polarisation. The rhombs are changed to inflated barrel-shaped prisms.

§ 17. Didymium

   a. Precipitation with sodium sulphate. Limit: 0.07 μgr. of Di.

b. Precipitation with sodium carbonate. Limit: 0.1 μgr. of Di.

c. Precipitation with oxalic acid. Limit: 0.1 μgr. of Di.

d. Precipitation with potassium ferrocyanide. Limit: 0.15 μgr. of Di.

a. The precipitation is even more slow than with lanthanum sulphate; the effect of heating also is less marked, otherwise no difference is noticed.

b. The statement found in Würtz's Dictionn. de Chimie, art. "Didyme," that didymium carbonate ($DiCO_3 + H_2O$) cannot be made to crystallise, is not quite correct. It certainly takes much more time than lanthanum carbonate, about half an hour, for crystallisation to commence, and after two hours only crystalline spherulites (20 to 40 μ) are found, but no needles or well-defined plates.

c. The precipitate with oxalic acid resembles cerous oxalate. It is less dense and changes more speedily to cruciform crystals and prickly spherulites.

d. A minute addition of potassium ferrocyanide yields similar grains as in solutions of cerous compounds (§ 15, *d*). With an excess of the reagent, after a minute, hexagonal plates are produced ($KDiFeCy_6 + 4H_2O$), sometimes with concave edges, while some of them will grow to elegant radiated and pinnated rosettes. When seen sideways these growths resemble round boxes with convex bottom and cover; if the terminals are unequal, they sometimes remind one of shirt-buttons. The

FIG. 22.—Double ferrocyanide of potassium and didymium, × 120.

diameter of these box-like or button-like structures may

reach 60 μ, their tint is a peculiar grayish-violet, characteristic of didymium. This reaction can be applied to the double sulphates of didymium; it is in this case slower than with the nitrate.

An approximate separation of lanthanum and didymium may be effected, if about 0.1 gr. of the mixed nitrates can be obtained. These are cautiously fused and kept for two minutes at the melting-point of lead. Water will leave a grayish basic nitrate of didymium. If the material be not too scanty, the operation should be repeated a second time.

§ 18. Yttrium and Erbium

a. Precipitation with oxalic acid.[1] Limit: 5 μgr. of Y.

b. Crystallisation of the oxalate from ammonium solution. Limit: 0.03 μgr. of Y.

a. The crystals of yttrium oxalate ($YC_2O_4 + 3H_2O$), precipitated from a neutral solution, are exceedingly small. Cruciform crystals, measuring 25 μ, can be obtained from solutions strongly acidulated with hydrochloric acid. The delicacy is, however, under these circumstances reduced in such a measure that the limit sinks to 5 μgr.

b. Yttrium oxalate dissolves sparingly in a solution of ammonium carbonate. If ammonium carbonate is added to a solution of a salt of yttrium and erbium till the carbonates are redissolved, a grain of oxalic acid will throw down a crystalline powder. By repeated heating (adding water and ammonium carbonate if necessary) octahedral polarising crystals (10 to 25 μ) can be produced, resembling crystals of calcium oxalate. A reaction by which yttrium can be distinguished from erbium is as yet wanting.

[1] Haushofer, l.c. p. 149.

§ 19. Barium

a. Precipitation with sulphuric acid. Limit: 0.046 μgr. of Ba.[1]

b. Precipitation with ammonium fluosilicate. Limit: 0.09 μgr. of Ba.[2]

c. Precipitation as chromate. Limit: 0.08 μgr. of Ba.[3]

d. Precipitation with tartrate of antimonyl and potassium. Limit: 0.45 μgr. of Ba.[4]

Besides these, potassium ferrocyanide,[5] oxalic acid,[6] tartaric acid,[7] and ammonium carbonate[8] have been proposed as reagents for tracing barium.

a. Barium sulphate dissolves in hot sulphuric acid, crystallising on cooling in rectangular plates (5 to 12 μ), and in rudimentary forms like the letter X, both belonging to the orthorhombic system. Calcium sulphate may be removed beforehand by treatment with hot water.

FIG. 23.—Sulphate of barium, × 300.

It is advisable to evaporate the greater part of the solvent, which is done with more ease and safety on a cover-glass than on a slide. If no crystals are seen after cooling, the preparation is cautiously moistened by breathing upon it. If a solution of barium chloride contains a great quantity of ferric chloride and hydrochloric acid, sulphuric acid will not precipitate pure barium sulphate. The precipitation is very slowly accom-

---

[1] Behrens, Mikr. Meth. p. 27; Haushofer, l.c. p. 16. [2] Bořický, l.c. p. 21.
[3] Haushofer, l.c. p. 17. [4] Streng, Ber. d. oberh. Ges. xxiv, p. 55.
[5] Streng, N. Jahrb. f. Miner., 1885, i. p. 39.
[6] Streng, l.c.; Haushofer, l.c. p. 19. [7] Streng, N. Jahrb., 1886, i. p. 56.
[8] Haushofer, l.c. p. 19.

plished; it is sometimes not terminated after two hours; the crystals have a yellowish tint and may grow to 50 μ. Chromic chloride and aluminium chloride will produce similar effects.

b. Instead of hydrofluosilicic acid, prescribed by Bořický, ammonium fluosilicate may be employed. It is easily purified, and will not corrode glass, if kept dry. Barium is rapidly and completely precipitated by this reagent from slightly acidulated solutions. The crystals of barium fluosilicate ($BaSiF_6$) are well developed rods (40 to 70 μ), derived, according to Streng,[1] from a rhombohedron. A sufficient quantity of the reagent and of acetic acid must be employed. Rapid heating is very serviceable, hastening the precipitation and fully developing the crystals. After this treatment, the form and arrangement of the crystals agree nearly with a crystallisation of calcium sulphate. Strontium and calcium are not precipitated by ammonium fluosilicate.

Fig. 24.—Fluosilicate of barium, ×60.

c. Haushofer prescribes neutral potassium chromate for the precipitation of barium compounds. Yet, with this reagent, neither good crystals are obtained nor a characteristic test, strontium being precipitated with barium. Potassium bichromate answers better, especially in solutions containing free acetic acid. Strong acids are made harmless by adding acetate of sodium or of ammonium; solutions of uncertain reaction are adjusted by adding acetate and acetic acid. Under these circumstances potassium bichromate will precipitate light yellow, square, and rod-like

[1] Streng, Anl. z. Best. d. Miner. p. 77.

crystals (8 to 20 μ) of the compound BaCrO₄. The precipitation is not complete at the end of half an hour; it may, however, be hastened by a slight heating, without any chance of precipitating strontium, if enough acetic acid has been added. To precipitate strontium a great excess of acetate must be added, and the heat must be raised to ebullition. Calcium is not precipitated, even by neutral chromate.

d. If tartar emetic be added to hot neutral solutions of barium salts, thin rhombic or six-sided plates of a double

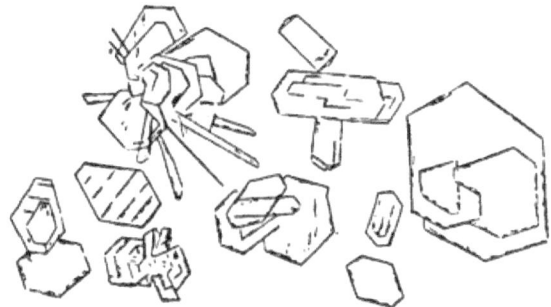

FIG. 25.—Stibio-tartrate of barium, ×90.

tartrate of antimonyl and barium $(Ba(SbO)_2C_8H_8O_{12} + 2H_2O)$ separate out on cooling. The following conditions must be strictly observed; the solution should be neutral or acidulated with acetic acid; it must not contain strontium, as this would be precipitated with barium; neither should calcium, magnesium, and alkali salts be present in great quantity, considering that the solubility (1 : 243) of the double tartrate will oblige the operator to concentrate poor solutions almost to dryness.

With regard to *potassium ferrocyanide* and *tartaric* acid, it is to be noted that they are neither characteristic reagents nor of special delicacy. If ammonium chloride be present, potassium ferrocyanide will precipitate calcium from highly diluted solutions, while the limit of dilution for barium is very low (1:38). By tartaric acid, strontium is precipitated as well as barium, interfering with the crystallisation of the tartrate of barium. *Oxalic* acid may be utilised for separating barium from strontium and calcium, if it be added to a solution acidulated with a moderate quantity of nitric acid. Oxalates of strontium and calcium come down immediately; barium oxalate separates out during evaporation, or after addition of sodium acetate, in slender prisms or in rhombs, measuring 20 μ. Boric acid retards the precipitation and reduces the prisms to curved and geniculated threads; aluminium chloride and ferric chloride have likewise a retarding action and will produce thread-like growths, but instead of short isolated threads long tufts (300 μ) are produced, resembling curled hair. The phenomenon is very striking, especially with iron (see § 43, *c*), and scarcely influenced by strontium and calcium.

Ammonium carbonate precipitates from pure solutions of barium salts characteristic stars (40 μ); if strontium, or calcium, or magnesium are present the reagent is of no use; it will under these circumstances precipitate insignificant globules (20 μ). Crystals of *barium nitrate* are sometimes accidentally produced, as this compound is but sparingly soluble in strong nitric acid. They are clear, well-defined octahedra, resembling caesium alum, freely soluble in water.

§ 20. Strontium

    a. Precipitation with sulphuric acid. Limit: 0.2 μgr. of Sr.[1]
    b. Precipitation as chromate. Limit: 0.8 μgr. of Sr.
    c. Precipitation as tartrate. Limit: 0.4 μgr. of Sr.[2]

[1] Behrens, Mikr. Meth. p. 26.   [2] Haushofer, l.c. p. 85.

d. Precipitation as carbonate. Limit: 0.4 μgr. of Sr.[1]

Besides these tests, precipitation with tartar emetic [2] and with oxalic acid [3] have been proposed.

a. From solutions in hot sulphuric acid strontium sulphate crystallises in rhombs and cruciform structures four times as large as the small crystals of barium sulphate. This reaction is of more value for strontium than for barium, because in a mixture of equal parts strontium will mask barium. Ferric chloride has a similar retarding action, as described in § 19, *a*. Pure hydrochloric acid has no sensible action on barium sulphate; heated with strontium sulphate, it will dissolve a considerable quantity of this compound. Strontium sulphate is decomposed under this treatment; if a large drop of hydrochloric acid is employed no crystallisation will take place on cooling, unless a trace of sulphuric acid be added, when crystals will separate out, reproducing various forms of celestine. After repeated treatment with hot hydrochloric acid, barium sulphate remains in a state of comparative purity. See § 83, Part II.

FIG. 26.—Sulphate of strontium, ×200.

b. Even bichromate of potassium will precipitate strontium chromate if a sufficient quantity of sodium acetate be added, and time allowed for the slow progress of the reaction at ordinary temperatures. It will commence at ordinary temperatures after half an hour; at a moderate heat, after a minute. Strontium chromate has nothing in common with barium chromate except its yellow colour. If it separates out from acidulated solutions at ordinary temperature, it will form dumb-bells and stout rounded

---

[1] Haushofer, l.c. p. 123.   [2] Streng, Ber. d. oberh. Ges. xxiv. p. 55.
[3] Streng, N. Jahrb. f. Miner., 1885, i. p. 38.

rods (length 20, diameter 8 to 12 μ). Usually highly refractive yellow globules (20 to 30 μ) are produced, growing to a considerable size (100 μ) under favourable circumstances. These globules are characteristic of strontium if zinc be absent. After precipitation of the greater part of barium (§ 19, c) the preparation is touched with a platinum wire dipped in caustic ammonia. A small cloud of a flaky or powdery precipitate, produced by local supersaturation, is of no consequence. It is soon redissolved, while at a small distance globules of strontium chromate make their appearance.

c. Tartar emetic will produce crystals resembling in all respects those described in § 19, d.

Neutral tartrate (Seignette salt) affords a test of far greater delicacy, not injured by acetic acid. The beautiful orthorhombic prisms of strontium tartrate will not be developed if barium is present; instead of crystals a powdery precipitate will be formed. Before trying the test with tartrate, barium must be precipitated as fluosilicate (§ 19, b). Another difficulty arises from the isomorphism of the tartrates of strontium and calcium. Numerous compound crystals are formed, and these exclude subsequent verification with sulphuric acid. Magnesium acetate will retard and modify the crystallisation of calcium tartrate, while it has no influence on the tartrate of strontium. If boric acid or soluble compounds of aluminium, chromium, or iron be present, the crystallisation of both tartrates is spoiled.

d. Sodium bicarbonate precipitates fibrous spheres (25 μ), composed of concentric layers. They are endowed with very strong double refraction. Between crossed nicols they will exhibit coloured circles up to the third order, intersected by a dark cross. Sodium bicarbonate has not yielded the sheaf-like and fan-like aggregates described by

Haushofer as having been precipitated with ammonium carbonate. In solutions containing barium, calcium, or magnesium, the form of strontium carbonate will be found predominant; it is, however, not advisable to employ sodium bicarbonate in testing such mixed solutions for small quantities of strontium.

e. Oxalic acid will precipitate solutions strongly acidulated with nitric acid. The clear sharp crystals (orthorhombic or monoclinic) have generally an octahedral aspect (20 to 40 μ); if much nitric acid has been added, or if aluminium, chromium, or iron be present in considerable quantities, a rectangular prism is developed, capped by pyramids. Calcium oxalate has the same octahedral aspect, but its crystals are half the size of the former (12 to 20 μ), and are not modified by nitric acid, nor by soluble compounds of aluminium or iron.

FIG. 27.—Oxalate of strontium, ×130.

The peculiar behaviour of stannic chloride towards the oxalates of calcium, strontium, and barium is worthy of notice. It may be referred to in this place, because the most striking effect is observed with strontium oxalate. If this reagent be added to a solution containing calcium, strontium, and barium, the order in which these metals are precipitated by oxalic acid is changed, especially if the solution has been slightly acidulated with hydrochloric acid. Strontium oxalate will separate out first, forming big octahedral crystals (60 to 100 μ) containing a little tin and exhibiting feeble polarisation. They seem to belong to the tetragonal system. If much acid and stannic chloride is employed, the crystals are corroded by the acid, and after some time they will have disappeared. If much barium is present, they are covered with prickles. Barium oxalate will appear next, forming crosses and six-rayed stars (40 to 120 μ). Calcium oxalate is the last to appear, yielding small squares and roundish grains (10 to 25 μ). By neutralising with small quantities of magnesium carbonate these crystallisations may be made to succeed each other. The limits are: 0.8 μgr. for Sr; 0.5 μgr. for Ba; 0.05 μgr. for Ca.

§ 21. Calcium

a. Precipitation as sulphate. Limit: 0.04 μgr. of Ca.[1]
b. Precipitation as tartrate. Limit: 0.03 μgr. of Ca.[2]
c. Precipitation with potassium ferrocyanide. Limit: 0.015 μgr. of Ca.
d. Precipitation with oxalic acid. Limit: 0.06 μgr. of Ca.[3]

Fluosilicate of calcium has been proposed by Bořický,[4] carbonate of calcium by Haushofer.[5]

a. Precipitation with sulphuric acid or with alkali sulphate has become of current use in examining minerals and rocks. This test, indeed, leaves little to be desired as regards delicacy, sureness, and neatness. From a solution in concentrated sulphuric acid short orthorhombic prisms of anhydrous calcium sulphate separate out on cooling. They are sometimes produced accidentally in testing for barium and strontium. Addition of sulphuric acid or of soluble sulphates to solutions of calcium salts will always lead to the crystallisation of the compound $CaSO_4 + 2H_2O$. From neutral solutions or from such as are but slightly

---

[1] Behrens, Mikr. Meth. p. 21.   [2] Haushofer, l.c. p. 85.
[3] Streng, N. Jahrb., 1885, i. p. 38.
[4] Bořický, l.c. p. 20.   [5] Haushofer, l.c. p. 38.

acidulated, it crystallises in slender monoclinic prisms (15 to 90 μ), terminated obliquely by the hemipyramid (65° 36'). They agree in all respects with the prismatic crystals of gypsum, like it forming swallow-tail twins. Sometimes double twins are seen resembling in shape an X. Strong acids diminish the delicacy of this test and lead to the crystallisation of needles grouped in sheaf-like or fan-like aggregates. The normal crystallisation is restored by acetate of sodium or acetate of ammonium. Chlorides of trivalent metals have a very injurious influence, retarding the crystallisation of calcium sulphate, and reducing its crystals to small squares and grains (10 to 15 μ) that cannot be distinguished from strontium sulphate. A remedy is found in heating the diluted sample with ammonium acetate and concentrating the clear solution. On the whole, testing with sulphuric acid for calcium will succeed best, if the sample be sufficiently diluted to allow some concentration after addition of the reagent. Under this treatment crystallisation proceeds from

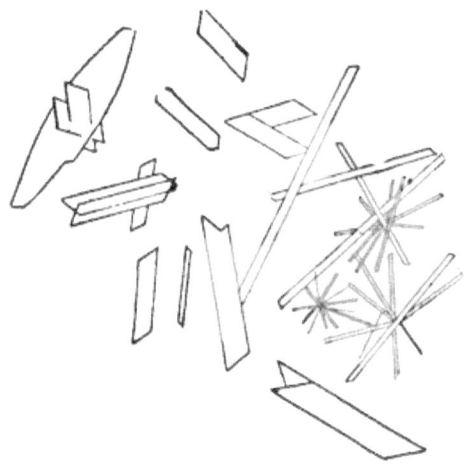

FIG. 28.—Sulphate of calcium, ×200.

the periphery to the centre of the drop, and will generally yield an ample crop of well-developed crystals. The presence of great quantities of alkaline salts will sometimes render this plan impracticable. For this reason, testing for calcium ought not to be deferred to a late stage of the examination. In very dilute samples calcium sulphate may be made to crystallise by covering the slide for about five minutes with a small box or bell-glass wetted with alcohol. Boric acid spoils the crystallisation in the same way as aluminium chloride; in this case no other remedy is known than to precipitate oxalate or carbonate of calcium.

b. If an acid solution has to be tested, add acetate of sodium or ammonium. If the solution is neutral, add a small drop of acetic acid. Testing with Seignette salt is slower than testing with sulphuric acid. This peculiarity is in favour of concentrated solutions, which will yield crystals of tartrate quite as perfect as those obtained from dilute solutions. With dilute solutions the action may be so slow as to make the operation tedious. It should be borne in mind that the great orthorhombic crystals will appear near the spot where the reagent has been put in. Usually the prism is predominant. Sometimes the facets of the dome will attain a greater extension, producing the semblance of an octahedron.

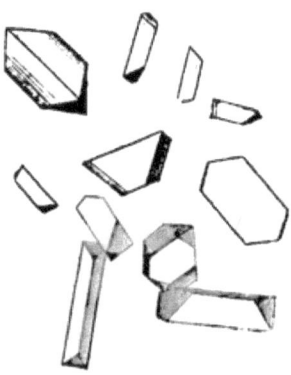

FIG. 29.—Tartrate of calcium, ×120.

Calcium sulphate is slowly dissolved by a solution of Seignette salt, stout crystals of the tartrate coming in the place of the slender rods and needles of the sulphate. On the other hand, the tartrate is easily decomposed by dilute sulphuric acid. Unfortunately the value of this beautiful reaction is impaired by several unfavourable circumstances; in the first place, by the isomorphism of the tartrates of strontium and calcium. This difficulty can be partially overcome by adding magnesium acetate, which will retard the crystallisation of calcium tartrate and will give its crystals a more slender, rod-like shape. Further, if barium is present, a powdery precipitate will be formed instead of crystals. Boric acid has a similar effect; a fine dust is thrown down, as if an acidulated solution of ammonium molybdate were evaporated. The injurious effect of the chlorides of aluminium, chromium, and iron closely resembles that of boric acid.

c. A feeble dose of potassium ferrocyanide has no perceptible effect upon solutions of calcium salts. An excess of the reagent precipitates small squares (20 to 25 μ) of potassium-calcium ferrocyanide ($K_2CaFeCy_6 + 3H_2O$). If ammonium chloride is present this test can be employed for extremely diluted solutions. Strontium is not precipitated; barium is precipitated from concentrated solutions (1 : 38), yielding pale yellow rhombohedra.

d. Calcium oxalate forms small tetragonal octahedra (12 to 20 μ). Their small size is an inconvenience which can be partially overcome by retarding the precipitation with nitric acid. It is not injured by boric acid nor by the chlorides of aluminium, chromium, or iron.

e. Calcium carbonate may be precipitated in an amorphous state by sodium carbonate. The amorphous grains are speedily changed to minute rhombohedra. If sodium bicarbonate or ammonium carbonate are employed, dilute solutions of calcium compounds will remain clear for some minutes. A slight heating will start the crystallisation of small rhombohedra (6 to 10 μ), which may grow to 25 μ. However small, their crystalline form is always developed with perfect neatness. It is not influenced by soluble compounds of aluminium, chromium, or iron, nor by boric acid, but it is spoiled by the presence of magnesium, strontium, or barium.

A double carbonate of calcium and sodium is sometimes produced if a great excess of sodium carbonate be employed in precipitating calcium carbonate. At first amorphous carbonate of calcium is thrown down. After some minutes the muddy appearance of the drop is seen to clear up, small stout monoclinic crystals coming in the place of the amorphous precipitate. They agree in form and composition with Gaylussite ($Na_2Ca(CO_3)_2 + 5H_2O$). After a superficial inspection they might be confounded with crystals of calcium sulphate.

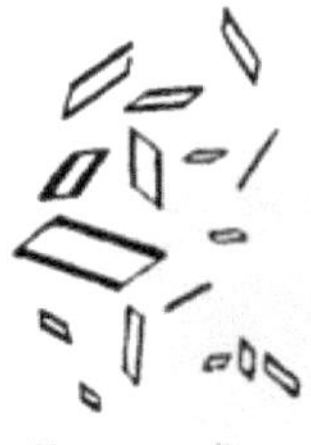

FIG. 30.—Gaylussite, ×130.

f. The fluosilicate of calcium is unfit for microchemical use; it is highly soluble, and will generally yield rudimentary crystals of uninviting appearance.

§ 22. Lead

a. Precipitation as chloride. Limit: 0.3 μgr. of Pb.[1]
b. Precipitation as iodide. Limit: 0.2 μgr. of Pb.[2]
c. Precipitation as sulphate. Limit: 0.04 μgr. of Pb.[3]
d. Precipitation as carbonate. Limit: 0.06 μgr. of Pb.
e. Precipitation as triple nitrite. Limit: 0.03 μgr. of Pb.

Haushofer has proposed, besides these tests, precipitation as oxalate[4] and as chromate.[5]

a. Lead chloride will on many occasions turn up accidentally in the course of microchemical operations. If hydrochloric acid be employed, it must be borne in mind that nitric acid may be set free, dissolving lead chloride in much greater proportion than water. Under such cir-

---

[1] Behrens, Mikr. Meth. p. 34.   [2] Haushofer, l.c. p. 28.
[3] Ibid. p. 26.   [4] Ibid. p. 29.   [5] Ibid. p. 28.

cumstances it is advisable to evaporate and to recrystallise from hot water. The crystals are slender prisms of the orthorhombic system, strongly polarising; from dilute hydrochloric acid thin rhombic plates; from solutions containing potassium chloride irregular bundles of thin threads. Ammonium chloride is more injurious than potassium chloride. With the chlorides of bismuth, antimony, and tin, lead chloride unites to form insoluble compounds, difficult to dissolve in dilute hydrochloric acid.

FIG. 31.—Chloride of lead, ×60.

b. Precipitation with potassium iodide yields six-sided yellow plates, glittering in reflected light with various tints.

FIG. 32.—Iodide of lead, ×90.

They are about half the size of the rhombic plates of lead chloride. An excess of potassium iodide is injurious, a colourless double iodide of greater solubility being formed. Lead iodide dissolves freely in hot water, recrystallising in great glittering scales and six-sided plates. Both are found to be dichroic, thus not hexagonal, but probably orthorhombic, like the chloride. Hot hydrochloric acid will dissolve a great quantity of lead iodide. On cooling some iodide is recovered. The greater part proves to have been decomposed. From acid solutions, containing iodides of lead, bismuth, and antimony, water will precipitate an orange-coloured or minium-coloured powder, containing lead, bismuth, and antimony. Hot water will dissolve a part of the lead iodide, which crystallises on cooling.

c. Sulphate of lead, precipitated with sulphuric acid or with an alkaline sulphate, has the appearance of a dense white powder of very fine grain. If it has been produced by decomposing sulphides containing lead by strong nitric acid, its appearance is more crystalline. Haushofer advises to crystallise it from hot sulphuric acid under the bell-glass

FIG. 33.—Sulphate of lead, × 300.

of a desiccator. A more simple means to the same end is heating it with nitric or hydrochloric acid. It dissolves freely under temporary decomposition, and may be made to crystallise partly by cooling, wholly by evaporation. The crystals are six-sided plates and rhombs (6 to 10 μ), isomorphous with barium sulphate. Lead sulphate is easily separated from sulphates of barium and strontium by treating the mixed sulphates with a strong solution of caustic alkali, in which lead sulphate dissolves readily. From this solution sodium bicarbonate will precipitate crystalline carbonate of lead.

d. The carbonate forms very characteristic branched rods (12 to 20 μ). It is precipitated immediately by sodium bicarbonate, and after about two minutes by ammonium carbonate.

e. The reaction to be described in § 23, *a*, treating of the tests for copper, may be modified so as to afford an excellent test for lead. If to a nearly neutral solution containing a small quantity of lead are successively added a small quantity of copper nitrate or acetate, sodium acetate, acetic acid, and an excess of potassium nitrite, the black cubes described in § 23, *a*, will very speedily appear. The delicacy of the test is about half that of the sulphate. If, finally, a grain of cæsium chloride be added, the delicacy of the sulphate is surpassed. If thallous nitrate be added, the triple nitrite will rival in delicacy the test with sulphuretted hydrogen. In the latter case the size of the crystals is reduced to 3 μ. They are, however, easily recognised by their dark brown colour and by their sharp outlines.

The *oxalate* of lead forms grains, small crosses (20 to 30 μ), or, if precipitated from hot solutions, rectangular rods (40 to 60 μ). The limit is about 0.2 μgr. of Pb. It is of little value for tracing lead.

Chromate of lead is precipitated from neutral solutions as a yellow powder, looking black in transmitted light. Bourgeois has shown that it can be made to crystallise by precipitating it from hot solutions, acidulated with nitric acid (limit: 0.08 to 0.15 μgr. of Pb). If the quantity of powdery precipitate be not too small, it may be transformed into orange-coloured crystalline basic chromate by adding a minute grain of caustic alkali. Both methods are of more importance for tracing chromium than for lead. For more detail see Chromium, § 44, *b*; and Ferrochrome, in Part II., § 121, *b*.

FIG. 34.—Chromate of lead, ×60.

Finally, mention must be made of the colourless, highly refracting octahedra of *lead nitrate*, often incidentally produced in dissolving alloys, ores and other compounds rich in lead. It is, like barium nitrate, sparingly soluble in nitric acid, but will dissolve freely in water. If it can be done, the acid solution should be run off from the octahedra, thereby eliminating a great quantity of lead and materially simplifying further examination.

§ 23. Copper

a. Precipitation with potassium nitrite and acetate of lead. Limit: 0.03 μgr. of Cu.

b. Precipitation with potassium ferrocyanide. Limit: 0.1 μgr. of Cu.[1]

c. Precipitation with ammonium thiocyanate. Limit: 0.1 μgr. of Cu.

d. Precipitation with potassium iodide and sodium sulphite. Limit: 0.05 μgr. of Cu.

a. Add a little sodium acetate to the drop that has to be tested, add a drop of a saturated solution of potassium nitrite, acidulate with acetic acid, and finally add a grain of lead acetate. As a considerable quantity of nitrite is essential, and as the product of the reaction is far from being insoluble, it is necessary to work with solutions as concentrated as possible. To attain the limit given above, the sample is evaporated and the residue is dissolved in the solution of nitrite or in the acetic acid. From the acetate of lead a yellow cloud is seen to spread in the liquid; shortly after a fine dust is perceived, and after about a minute black cubes (10 to 25 μ) are formed. Very thin plates of this compound are transparent, dark orange or brown; but their colour is of such an intensity that crystals of 3 μ look black. On analysis the composition of this remarkable compound has been found to agree

FIG. 35.—Triple nitrite of copper, lead, and potassium, ×60.

[1] Haushofer, Sitzungsber. d. bair., Akad., 1885, p. 412.

with the formula $K_2CuPb(NO_2)_6$, analogous to the triple nitrite of nickel (§ 12, a). If enough potassium nitrite be employed, the triple nitrite may be dissolved by heating it with the mother liquor and then recrystallised by slow cooling to a size of 70 μ. By ammonia it is bleached, by acetic acid the black tint is restored. Barium and strontium may be used in the place of lead, but the reaction is slower and the size of the crystals is smaller. With cæsium in the place of potassium the delicacy is trebled, but the size of the crystals is halved. With thallium the limit lies far beyond 0.01 μgr. of Cu; the size of the crystals is reduced to 3 μ.

b. Dissolve in a considerable excess of ammonia, add a grain of potassium ferrocyanide. Upon the evaporation of the ammonia a crystalline pale yellow compound [$(N_2H_6Cu)_2FeCy_6 + H_2O$] will separate out, taking slowly a reddish tint. Well-developed rhombs and squares are rarely observed; but plume-like forms will generally appear. Yet even these may be utilised for a beautiful and characteristic reaction; a drop of acetic acid will change their pale colour to blood-red.

c. The crystals of cupric-mercuric thiocyanate ($HgCy_2S_2 \cdot CuCy_2S_2 + H_2O$) are yellowish-green and more slender and pointed than those of the analogous compound of cobalt. If both copper and cobalt are present, the blue crystals of the cobaltous compound will appear first. The test will succeed best in a solution containing free acetic acid. It is characteristic but slow. A most curious modification is seen if copper is associated with a much greater quantity of zinc. Black spheroids and irregular black lumps will separate out, having nothing in common with cupric-mercuric thiocyanate, nor with the analogous compound of zinc. If only a trace of copper be present the form of zinc-mercuric thiocyanate is restored, but its crystals have a chocolate colour. The same phenomenon is observed with cadmium instead of zinc; it is, however, less marked. If, at the same time, a trace of cobalt be present, the crystals will exhibit a brownish-violet tint. It is probable that in a great excess of the anhydrous compound $HgCy_2S_2 . ZnCy_2S_2$ the small quantity of the analogous cupric compound is incorporated as an isomorphous anhydride, which is not known in an isolated state. This modified test is of great delicacy for tracing copper, rivalling in this respect § 23, *a*.

FIG. 36.—Double thiocyanate of copper and mercury, ×60.

d. Precipitation as cuprous iodide, Cu_2I_2, affords a delicate but not a convenient test, considering the fine grain and absence of colour. Concentrated solutions of cupric salts are precipitated by potassium iodide, iodine being set free; dilute solutions require addition of sulphuric acid and of sodium sulphite. The precipitate is a white crystalline powder, dissolving readily in ammonia, but not developing well-defined crystals with this solvent.

§ 24 and § 25. Mercury

(1) *Mercurous Compounds*

a. Precipitation with hydrochloric acid. Limit: 0.25 μgr. of Hg.[1]

b. Precipitation with potassium bichromate. Limit: 0.5 μgr. of Hg.

a. Hydrochloric acid will precipitate slender needles resembling needles of gypsum, precipitated from strongly acidulated solutions. In less than a minute these needles will crumble into grains of such small size that a power of 600 is required for ascertaining that they belong to the tetragonal system. In the case of highly-diluted solutions addition of ammonia is useful. The black colour produced by ammonia will generally suffice for tracing mercurous compounds. It is readily seen in minute particles if a low power and strong reflected light are used.

b. Potassium bichromate produces a precipitate of a fiery red. If a solution, containing free nitric acid, is precipitated while hot, red cruciform crystals (10 to 25 μ) are formed. An excess of the reagent will produce much larger rhombic crystals from yellow to orange, containing both mercury and potassium.

FIG. 37.—Mercurous chromate, ×130.

---

[1] Behrens, Mikr. Meth. p. 34.

(2) *Mercuric Compounds*

a. Precipitation with potassium iodide. Limit: 0.075 μgr. of Hg.[1]

b. Precipitation with stannous chloride. Limit: 0.05 μgr. of Hg.

c. Precipitation with ammonium thiocyanate and cobaltous acetate. Limit: 0.04 μgr. of Hg.

a. Potassium iodide produces in solutions of mercuric salts a crimson precipitate of mercuric iodide (HgI_2), soluble in hot water. From this solution it crystallises in tetragonal plates and pyramids of a ruby colour. Small crystals look black in transmitted light. A rapid precipitation will sometimes yield slender yellow needles of the monoclinic modification, which sooner or

FIG. 38.—Mercuric iodide, × 130.

later turn red. An excess of potassium iodide dissolves mercuric iodide. From the colourless solution it can be reprecipitated with cupric sulphate, with a tint even more vivid, unless a great excess of potassium iodide had been employed.

b. Stannous chloride precipitates mercurous chloride (§ 24, *a*). This test is delicate, but not convenient.

c. A test, highly to be recommended for its elegance and delicacy, is furnished by inverting the test for cobalt

---

[1] Haushofer, l.c. p. 112.

described in § 11, *b*. If ammonium thiocyanate and a grain of cobaltous acetate are added to a solution of a mercuric compound, blue crystals of cobaltous-mercuric thiocyanate (§ 11, *b*) are produced. If the liquid contain zinc, the crystals are light blue. Addition of zinc is useful in the case of highly diluted samples. An excess of reagent will do little harm. This test leaves little to be desired, only bismuth and lead need to be removed before testing.

§ 26. Gold

a. Precipitation with stannous and stannic chlorides. Limit: 2 μgr. of Au.

b. Precipitation with thallous nitrate. Limit: 6 μgr. of Au.

In addition to these means of tracing gold, its behaviour towards ammonium thiocyanate may be utilised.

a. The fine grained red precipitate, produced in solutions of gold by mixtures of stannous and stannic chlorides, is well fitted for microscopical observation, much better than the blue precipitate produced by oxalic acid, or the blackish precipitate produced by mercurous nitrate. It answers best to work with two drops, which are made to touch; where they unite, a well-marked red line will be perceived.

b. Instead of the uncertain crystallisation of chloride of gold, proposed by Haushofer,[1] I would direct the attention of chemists to the beautiful reaction, which is afforded by thallous chloroaurate ($TlAuCl_4 + 5H_2O$). If a grain of thallous nitrate is put into a solution of gold chloride (about 0.1 per cent of Au), bright yellow needles are seen sprouting from the thallous nitrate, and growing to a length

---

[1] Haushofer, l.c. p. 50.

of 100 μ. The affinity of thallium for gold is so strong, that thallous chloride is dissolved by gold chloride in the presence of hydrochloric acid. Heat is advantageous in this experiment, but it must be cautiously applied, as too much heat would lead to reduction of gold and the production of thallic chloride. This latter compound exhibits pale grayish hexagons, mingling with the yellow needles of the chloroaurate. The caution applies especially to dilute solutions, requiring concentration after addition of thallous nitrate. If the concentration is managed with prudence, the yellow needles will grow from the periphery towards the centre of the drop. It may be noted, by the way, that mercuric chloride will yield with thallous nitrate colourless needles of the same form and dimensions.

FIG. 39.—Thallous chloroaurate, × 20.

c. With ammonium thiocyanate, gold chloride behaves like mercuric chloride. A small quantity of ammonium thiocyanate produces a red powdery and transparent precipitate. With an excess of the reagent this precipitate is transformed to woolly rosettes of a pale red colour, soluble in ammonia and freely dissolving in a hot solution of ammonium thiocyanate. The colourless solution precipitates solutions of zinc and cobalt in the same manner as a solution of ammonium-mercuric thiocyanate. Double thiocyanate of zinc and gold is bright yellow, double thiocyanate of cobalt and gold is dark blue-green; compound crystals of these double thiocyanates exhibit a sea-green tint. Their form agrees with that of the analogous mercuric compounds. Solutions of nickel are not precipitated, in conformity with the behaviour of ammonium-mercuric thiocyanate.

§ 27 and § 28. Platinum

(1) *Platinum Bichloride—Platinous Chloride*

a. Precipitation as chloroplatinite of cupridiammonium (salt of Millon and Commaille). Limit : 0.06 μgr. of Pt.

a. To the solution containing platinum bichloride add a small quantity of a cupric salt, a strong dose of ammonium chloride, and an excess of ammonia. Violet needles of an intense colour will speedily appear ($N_2H_4(NH_4)_2Cu . PtCl_4$), growing to a length of 200 μ in a solution containing 0.1 per cent of potassium chloroplatinite. From solutions containing 0.01 per cent of this compound, short needles separate out after some minutes, showing a distinct colour.

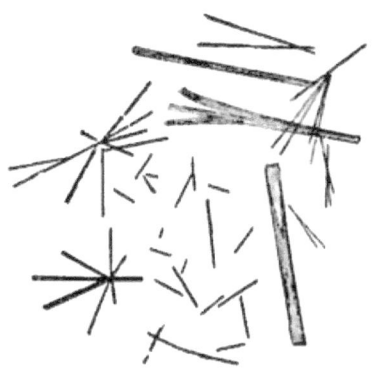

FIG. 40.—Chloroplatinite of cupridiammonium, ×90.

Palladous compounds will not yield an analogous reaction.

Solutions of moderate concentration can be tested by adding simply an excess of ammonia. Green needles of the salt of Magnus (chloroplatinite of platinodiammonium, $N_2H_4(NH_4)_2Pt . PtCl_4$) are produced.

(2) *Platinum Tetrachloride—Chloroplatinic Acid*

a. Precipitation with potassium chloride.   Limit: 0.6 μgr. of Pt.[1]

b. Precipitation with rubidium chloride.   Limit: 0.2 μgr. of Pt.

c. Precipitation with thallous nitrate.   Limit: 0.004 μgr. of Pt.

a. For the physical properties of potassium chloroplatinate see § 1, *a.* In dilute solutions the action of potassium chloride is slow. Solutions containing 0.1 per cent of platinum tetrachloride must be concentrated to less than half their volume. It will often happen that the drop is nearly dried up when crystals begin to appear. Yet even under such unfavourable circumstances the test is reliable; the octahedra will finally show themselves and will grow to good size and shape, unless the crystallisation is impaired by a great quantity of alkali salts.

b. Solutions containing 0.1 per cent of platinum tetrachloride are immediately precipitated by rubidium chloride. The octahedra are only one-third the size of the former, yet they are perfectly developed and easy to recognise. The action is slow in a solution of 0.03 per cent, the octahedra, produced after five minutes, very small. In this solution cæsium chloride shows instantaneous action, producing crystals of such small size that a power of 250 must be employed. Haushofer has cautioned against confounding chloroplatinate with chloropalladate and chloroiridate. Palladium tetrachloride is decomposed by heating with water; iridium tetrachloride produces a reddish colour in crystals of chloroplatinate, easily perceived on comparison with a standard

---

[1] Behrens, Mikr. Meth. p. 22.

slide. It can be eliminated by boiling with a solution of oxalic acid. Palladium is detected with potassium iodide.

c. Precipitation with thallous nitrate affords a test of great delicacy, at the same time rendering confusion with iridium impossible. The crystals are exceedingly small (§ 6, *c*); from concentrated solutions a yellow powder is thrown down. With some patience a good reaction may be got in a dilution of 1 : 20000. Iridium tetrachloride is reduced by thallous nitrate; a flaky compound of a dirty green colour being thrown down.

§ 29. Palladium

a. Precipitation as iodide. Limit: 0.1 μgr. of Pd.

b. Precipitation as thallous chloropalladite. Limit: 0.2 μgr. of Pd.

c. Precipitation with ammonium thiocyanate and thallous nitrate. Limit: 0.07 μgr. of Pd.

a. A feeble dose of potassium iodide yields a dark brown powdery precipitate, dissolving with a reddish-brown colour in an excess of the reagent. In solutions of platinum tetrachloride a dark wine colour is produced, and afterward dark gray octahedra of potassium iodoplatinate. With ammonia, palladium iodide yields small orange-coloured needles and a colourless solution, from which iodide of palladammonium ($N_2H_6PdI_2$) separates out in light yellow rectangular dendrites (20 to 30 μ). An excess of reagents, especially of potassium iodide, is

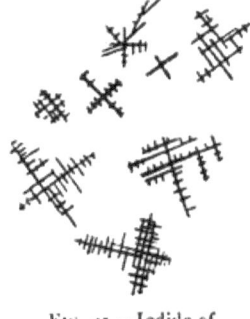

Fig. 41.—Iodide of palladammonium, ×130.

injurious. Heating with hydrochloric acid is to be employed as remedy.

b. With a solution of palladous chloride, thallous nitrate yields a reaction, agreeing in many respects with that described in § 26, *b*. The rods and needles of the palladium compound are light brown. In solutions of palladous nitrate the same reaction is obtained by adding a small drop of hydrochloric acid, and afterwards a grain of thallous nitrate. The compound may be dissolved by heating with water or with dilute hydrochloric acid, and recrystallised by concentration. Addition of potassium iodide gives a little more delicacy and crystals of smaller size.

FIG. 42.—Thallous chloropalladite, × 130.

A test of nearly the same delicacy, and one which is very useful for separations, is afforded by the feeble solubility of the *chloride of palladammonium* ($N_2H_6PdCl_2$). If ammonium chloride and an excess of ammonia are added to a solution of palladium, a colourless, freely soluble compound is formed, chloride of palladodiammonium ($N_2H_4(NH_4)_2PdCl_2$). If this be slightly heated with dilute hydrochloric acid, light brown, highly characteristic dendrites of the chloride of palladammonium will crystallise. See Part II., V. (4), § 148, Examination of Native Platinum.

c. Ammonium thiocyanate will not precipitate solutions of palladium. Addition of thallous nitrate will produce a brown precipitate, dissolving in hot water, and crystallising from this solution in rectangular dendrites (100 μ), which develop into glittering scales. From very dilute solutions rectangular prisms and crosses (tetragonal or orthorhombic) are produced (with platinum stout rhombs). The reaction takes place even when there is a dilution of 1 part of palladium nitrate in 5000 parts of water.

With mercuric cyanide and ammonia colourless cubes (10 μ) are produced in solutions of palladium. This reaction is characteristic, but not at all delicate. Its limit is 2 μgr. of Pd.

Towards calcium oxalate palladium nitrate exhibits the same behaviour as ferric chloride and platinum tetrachloride towards barium oxalate, producing tufts of curled light brown hairs (limit: 0.5 μgr. of Pd). Strontium oxalate is modified in a similar manner, barium oxalate yields long rods. With platinum tetrachloride calcium oxalate is modified to a brownish crystalline powder, strontium oxalate to short rods, barium oxalate to hair-like growths.

§ 30. Iridium

a. Precipitation with rubidium chloride. Limit: 0.3 μgr. of Ir.

a. Ammonium chloroiridate dissolves in water with a reddish tint. Potassium chloride precipitates slowly dark red octahedra (25 μ). A solution of ammonium chloroiridate in 300 parts of water is immediately precipitated by rubidium chloride; a solution in 1500 parts of water is precipitated after some minutes. Bright red octahedra (10 μ) are slowly formed. This diluted solution is immediately precipitated by cæsium chloride. The crystals are very small, but well defined.

§ 31. Rhodium

a. Precipitation with potassium nitrite. Limit: 0.09 μgr. of Rh.

b. Precipitation with acid potassium oxalate.   Limit : 0.4 μgr. of Rh.

a. Potassium nitrite (recommended by Gibbs) precipitates rhodium from acid and from ammoniacal solutions. The precipitate of potassium-rhodium nitrite resembles the analogous compound of cobalt (§ 11, *a*) ; it is composed of small yellow cubes (2 to 4 μ), adhering to the glass. The reaction appears with a dilution of 1 part of rhodium in 1000 parts of water. The delicacy of the test may be increased by employing cæsium chloride. It is somewhat diminished in ammoniacal solutions; but, on the other hand, these yield very characteristic six-leaved flower-like rosettes. A slight heating is favourable, both in acid and in ammoniacal solutions.

FIG. 43.—Double nitrite of cæsium and rhodium, × 300.

b. This reaction is neither delicate nor characteristic. The precipitate is composed of fine short needles, like the oxalate of cobalt. (Palladium yields long slender prisms.)

c. Potassium ferrocyanide precipitates rhodium from highly diluted solutions. The flaky yellowish precipitate turning to a rose tint with an excess of ammonia is not characteristic if other metals, precipitated by the same reagent, are present.

§ 32. Ruthenium

a. Precipitation with cæsium chloride.   Limit : 0.8 μgr. of Ru.

b. Colouring with thiocyanate of potassium. Limit: 1.2 μgr. of Ru.

a. The brown solution of ruthenium in aqua regia is precipitated by cæsium chloride. The reddish-brown precipitate dissolves in hot water; the solution will not yield well-developed crystals, only reddish-brown grains of 3 μ are formed. Thallous nitrate produces yellow grains of 4 to 5 μ.

b. The purple colour produced by soluble thiocyanates in solutions of ruthenium can be utilised for microchemical examination. The sample is concentrated as far as possible, and touched with a platinum wire, dipped in a saturated solution of potassium or ammonium thiocyanate. If the test does not succeed, the assay should be concentrated cautiously, and the border of the drop examined from time to time. This reaction is rendered useless by iron, cobalt, platinum, and palladium.

§ 33. Osmium

a. Precipitation with cæsium chloride. Limit: 0.1 μgr. of Os.

b. Reduction of osmate to osmite. Limit: 0.1 μgr. of Os.

a. Solutions of osmic acid in hydrochloric acid yield with cæsium chloride a whitish crystalline precipitate, composed of light greenish-yellow octahedra (10 to 30 μ). Their colour does not agree with that of the compound K_2OsCl_6, which is described as being brick-red.

b. Osmic acid dissolves in a solution of caustic potash with a light brown colour. By alcohol the osmic acid is reduced, violet octahedra (orthorhombic) of potassium osmite ($K_2OsO_4 + 2H_2O$), measuring 50 μ, will separate out. Dilute solutions may be heated with impunity for concentration if a considerable excess of caustic potash be added. The delicacy of the test may be doubled by adding a strong dose of ammonium chloride to the alkaline solution of osmite. It is then transformed to light yellow fringed rods and dendrites of the ammoniacal compound $\left. \begin{array}{c} Os \\ H_{12} \end{array} \right\} N_4Cl_2$,

FIG. 44.—Osmite of potassium, ×130.

nearly insoluble in a concentrated solution of ammonium chloride. After crystallisation of this compound the presence of osmium may be yet further established by adding potassium ferrocyanide, which will produce an intense purple tint. In all experiments with osmium tetroxide, it is necessary to guard against the reducing action of organic substances.

§ 34 and § 35. Tin

(1) *Stannous Chloride*

a. Precipitation with gold chloride. Limit: 0.07 μgr. of Sn.

b. Precipitation with mercuric chloride. Limit: 0.07 μgr. of Sn.

c. Precipitation with oxalic acid. Limit: 1.0 μgr. of Sn.[1]

For *a* and *b* see § 26, *a*, and § 25, *b*. With regard to *c*, it is to be noted that the use of potassium oxalate instead of oxalic acid will afford a test of far greater delicacy. The crystals of stannous oxalate (SnC_2O_4) remain usually in

[1] Haushofer, l.c. p. 154.

a rudimentary state, exhibiting the form of the letter H. They show strong double refraction, extinguishing at an angle of about 15°. If arsenic acid is present, much tin

FIG. 45.—Stannous oxalate, ×90.

may be concealed as insoluble arsenate, even if a great quantity of hydrochloric acid be employed.

(2) *Stannic Chloride*

a. Precipitation with cæsium chloride.   Limit: 0.45 μgr. of Sn.

a. As regards the properties of cæsium chlorostannate, see § 4, *a*. Sometimes it may be useful to stain the crystals light yellow by adding potassium iodide. If concentration of the sample be possible, and if not much hydrochloric acid has been added, rubidium chloride and even potassium chloride may be employed as reagents, producing crystals of greater size. The limit is reduced to 1 μgr. of Sn with rubidium chloride, to 2.5

FIG. 46.—Chlorostannate of cæsium, ×130.

μgr. of Sn with potassium chloride. Metastannic acid is brought within the range of this test by heating with strong hydrochloric acid.

b. The modification, produced by stannic chloride in the precipitation of strontium oxalate, has been described in § 20, *c.* It may be turned to profit for tracing tin. A grain of strontium acetate is added, the solution is slightly acidulated with hydrochloric acid; it is concentrated as far as possible and tested with a grain of oxalic acid. With strontium, the limit is 0.2 μgr. of Sn; with barium, the test is far less delicate. Metastannic acid must be made soluble by fusion with caustic potash. It may be brought to the test in another way: by reducing it with metallic magnesium, dissolving the tin in hydrochloric acid, and applying the test with gold chloride, § 34, *a.*

§ 36. Titanium

a. Precipitation as fluotitanate of potassium. Limit: 6 μgr. of Ti.[1]

b. Precipitation as fluotitanate of rubidium. Limit: 1 μgr. of Ti.

c. Staining of titanium dioxide with potassium ferrocyanide.

a. Fusion with sodium fluoride (or with a mixture of sodium carbonate and ammonium fluoride) is a convenient method for the decomposition of insoluble compounds of titanium. The product is treated with dilute hydrochloric acid on platinum or on a varnished slide. Sodium fluotitanate dissolves readily in water and in dilute acids. The proposal of Haushofer to employ this compound for estab-

---

[1] Haushofer, l.c. p. 131.

lishing the presence of titanium cannot be taken seriously. Addition of a soluble compound of potassium produces the less soluble compound $K_2TiF_6 + H_2O$, crystallising slowly in the form of rectangles, octagons, and long six-sided plates (60 to 90 μ). They are pale and show feeble polarisation.

FIG. 47.--Fluotitanate of rubidium, ×130.

b. Rubidium chloride produces crystals of the same form (15 to 40 μ) in a much shorter time. In solutions containing 0.2 per cent of TiO_2 the crystallisation commences after two minutes. Cæsium chloride immediately precipitates crystalline scales, measuring 8 to 12 μ.

c. Fusing with sodium carbonate must be done in a very hot flame, approaching a white heat, or a little caustic alkali must be added, when a bright red heat will suffice. The product is washed with water, which leaves an insoluble acid titanate of sodium, soluble in strong hydrochloric acid. This solution may be tested as above after addition of ammonium fluoride. Or titanium dioxide is precipitated by boiling the diluted solution, and tested by staining it with potassium ferrocyanide. Very impure samples are evaporated with a small drop of sulphuric acid. The residue is heated with a large drop of water, washed with dilute hydrochloric acid, and tested with potassium ferrocyanide, which stains titanium dioxide a vivid brown. For further particulars see Titaniferous Minerals, Part II. § 152, e; § 153, c.

§ 37. Zirconium

a. Precipitation with acid potassium oxalate. Limit: 0.06 μgr. of Zr.

b. Crystallisation as potassium fluozirconate. Limit: 5 μgr. of Zr.[1]

c. Precipitation as rubidium fluozirconate. Limit: 0.5 μgr. of Zr.

a. Solutions of zirconium sulphate or of zirconium chloride are precipitated by oxalic acid if the dilution does not exceed 1:200, and if only a small quantity of free acid is present. Acid potassium oxalate will precipitate solutions containing 0.01 per cent of zirconium sulphate. The precipitate, potassium-zirconium oxalate $(K_4Zr(C_2O_4)_4 + 4H_2O)$, is composed of small colourless crystals (20 to 60 μ) of octahedral appearance, belonging probably to the tetragonal system, and resembling strontium oxalate. They are easily dissolved by hydrochloric acid. Solutions of titanium dioxide are not precipitated by acid potassium oxalate unless sodium acetate be added. The precipitate is composed of small rods (15 μ). Zircon may be decomposed in the same way as titaniferous minerals; if fusion with sodium fluoride has been employed, the product is treated with sulphuric acid.

Fig. 48.—Double oxalate of potassium and zirconium, × 200.

b. Fusion with potassium fluoride yields a crystalline semi-vitreous mass. From its solution in hot water rectangular prisms (15 to 20 μ) will settle. They represent the normal fluozirconate K_2ZrF_6.

[1] Haushofer, l.c. p. 157.

REACTIONS

c. From solutions of zirconium sulphate or of zirconium chloride another fluozirconate is produced, probably K_3ZrF_7. The potassium compound is too soluble for microchemical use. Better results are obtained with rubidium chloride as a reagent. First, ammonium fluoride is added and, if necessary, a little hydrochloric acid, then a grain of rubidium chloride. A varnished slide must be used for this test. Only colourless highly refractive octahedra are produced (30 to 60 μ). Acid solutions of the mass described in *b* also yield octahedra with rubidium chloride; if the solution be very concentrated rectangular prisms are also formed. For the decomposition of zircon by fusion with 2 parts of sodium carbonate, a white heat and much time are needed.

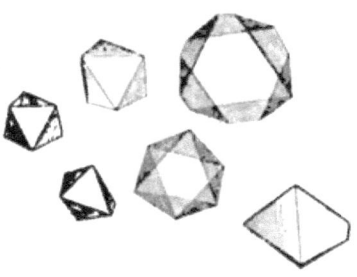

FIG. 49.—Fluozirconate of rubidium, ×200.

Lixiviation with hot water leaves a heavy powder, resolved under high powers into grains of sodium zirconate and into small hexagonal plates of zirconium dioxide, stained light yellow by a little platinum dioxide. This test, devised by Michel-Lévy and Bourgeois, is of sufficient delicacy; it will indeed give an unequivocal result with 5 μgr. of Zr, but it is not convenient, as a high power is necessary, and as the production of granular sodium zirconate cannot be avoided.

§ 38. Thorium

a. Precipitation of thorium sulphate by heat. Limit: 30 μgr. of Th.

b. Precipitation with oxalic acid. Limit: 0.1 μgr. of Th.[1]

[1] Haushofer, l.c. p. 128.

c. Precipitation with ammonium carbonate and thallous nitrate. Limit : 0.05 μgr. of Th.

a. Near one of the corners of a slide a cover glass is placed, and on this the drop that has to be tested. It must be heated rapidly to ebullition, breaking the crystalline crust round the drop with a platinum wire, and then pushing the cover glass towards the cold part of the slide. If a sufficient quantity of thorium be present, the drop will be full of slender needles (70 to 100 μ) of the sulphate $Th(SO_4)_2 + 3H_2O$. The sulphates of the cerite metals, especially the sulphate of lanthanum, exhibit the same phenomenon, although in a less striking manner. Confusion with them and with gypsum is avoided by adding ammonia and ammonium carbonate (see *c*) after the experiment. Calcium and the cerite metals are precipitated as carbonates, while thorium is dissolved.

b. Oxalic acid produces in solutions of thorium compounds a powdery precipitate, scarcely dissolving in hydrochloric acid, somewhat more soluble in ammonia and ammonium carbonate. Very dilute solutions yield a crystalline oxalate of very fine grain, composed, according to Haushofer, of square plates. I have not been able to obtain them otherwise than quite exceptionally; the majority of the small crystals had the appearance of minute rods (4 μ).

FIG. 50.—Double carbonate of thallium and thorium, ×300.

c. Thorium carbonate dissolves in ammonium carbonate, especially if caustic ammonia is added. A grain of thallous nitrate precipitates from this solution colourless rhombs and octahedral crystals of rhombic section. From very dilute solutions small rhombs (4 to 10 μ) separate out around the drop; in less dilute solutions the crystals will grow to a size of 40 μ. Beryllium, zirconium, and yttrium will not yield a reaction of this kind. Uranium yields a crystalline double carbonate under the same circumstances. It is easily distinguished from the double carbonate of thorium by its yellowish colour and by its solubility in oxalic acid.

§ 39. Silicon

a. Precipitation as sodium fluosilicate. Limit: 0.05 μgr. of Si.

b. Precipitation as rubidium silicomolybdate. Limit: 0.004 μgr. of Si.

a. Dissolve in hydrofluoric acid or in a mixture of ammonium fluoride and hydrochloric acid, and precipitate with sodium chloride. Sodium fluosilicate (§ 2, *c*) forms six-sided plates and stars (40 to 70 μ) belonging to the hexagonal system; from more concentrated solutions elegant six-spoked rosettes are produced. All these crystalline growths have sharp outlines and a peculiar faint rose tint. It is an unfortunate circumstance, militating against the use of this compound, that its form and tint recur with several fluosalts of sodium, viz. with the fluozirconate, the fluostannate, the fluotitanate, and the fluoborate. They are far more soluble than the fluosilicate. Separation may be effected before testing by distillation with ammonium fluoride and sulphuric acid; a drop of water or of a solution of ammonium acetate being employed to absorb the fluoride of silicon. If the heat be not raised above 140°, boron fluoride remains in the residue. Distillation may be avoided by employing potassium chloride as reagent. Potassium fluosilicate crystallises in colourless pale cubes (10 to 15 μ), much smaller than the hexagons of the sodium compound. If silicates are decomposed with hydrofluoric acid or with acid solutions of fluorides, it may happen that crystals of potassium or sodium fluosilicate are formed in the course of this treatment.

b. The sample is fused with five times its volume of sodium carbonate, the fused mass is dissolved in water, ammonium molybdate and an excess of nitric acid are added to the solution. If phosphoric acid be present, yellow grains of phosphomolybdate will be formed (§ 52, *b*). Phosphoric acid is eliminated together with arsenic acid at a moderate heat. The heat is then raised to ebullition, with a view to the formation of silicomolybdate of ammonium. After cooling, rubidium chloride is added to the clear liquid (see § 5, *a*). This test is exceedingly delicate, but just for this reason certain precautions must be observed. In the first place, it must be borne in mind that from an acid solution of ammonium molybdate small needles of an acid molybdate may separate out, and that it will leave on evaporation a fine-grained residue. Thus a granular precipitate must not be taken for a conclusive proof of the presence of silica. In the second place, the test is of sufficient delicacy to betray the minute quantity of silica, taken up from the glass, if the boiling has been done on a glass slide. For this reason, the acid liquid ought to be boiled and nearly evaporated on platinum, the residue ought to be heated with water, and the experiment is to be terminated by concentrating and testing the solution on a slide. Finally, it is to be noted that the acids of vanadium, niobium, and tantalum, and the oxide of bismuth may form complex molybdates of the same kind as arsenic molybdate; they are precipitated by potassium compounds, which is not the case with silicomolybdate. Thorium dioxide will form a thorium molybdate, agreeing in all respects with silicomolybdate; while the dioxides of zirconium, titanium, and tin form complex molybdates, differing essentially from silicomolybdate.

c. Haushofer has tried to develop staining of gelatinous silica with fuchsine (see Part II., § 92) into a microchemical test. It is not of sufficient delicacy for this purpose; besides, the stain, recommended by me for specimens of rocks, is of no use for gelatinous silica suspended in a liquid. Accidental films of the dye-stuff are often deposited on the glass when no trace of gelatinous silica is present, necessitating the washing in a great quantity of water for at least half an hour. Malachite green has, however, been found to stain gelatinous silica more effectually than fuchsine, and to be less apt to produce accidental films.

§ 40. Carbon

a. Precipitation as carbonate of lead or of strontium. Limit: 1 μgr. of C.

b. Expulsion of gaseous carbonic acid. For Cyanogen see § 53.

a. From solutions of carbonates, carbonic acid may be precipitated by calcium acetate. The powdery precipitate will soon be transformed to minute rhombohedra (§ 21, *e*). Lead carbonate crystallises more speedily and forms larger crystals. If rapidly precipitated, it exhibits small rods, twinned to the likeness of an X, and ramified modifications of these rods and twins. If precipitated slowly, it forms lenticular crystals (15 μ).

Graphite and carbon are burnt with nitre. If these substances are ground to fine powder with ten times their volume of nitre, the combustion can be brought about without sensible loss by heating on a platinum spatula. No more heat must be applied than strictly necessary. Even if the heating is managed with due caution, some nitrite is formed, excluding the use of lead acetate for reagent. Strontium acetate may be used in its place, yielding characteristic spheroids of carbonate (6 to 12 μ), described in § 20, d.

b. From insoluble carbonates carbonic acid is driven out by hydrochloric acid or by nitric acid, under a cover glass which arrests the escaping bubbles. The insoluble carbonate is immersed in a drop of water, covered, and brought into full view under a low power of the microscope. Close to the cover glass a drop of acid is placed, and on the opposite side a strip of dry filtering-paper, which starts the reaction by absorbing the water. Under this form the test has come into universal use for examining specimens of rocks. More delicacy and a more exact localisation of the bubbles is gained by spreading a thin layer of a mixture of gelatine and glycerine over the sample. The reaction is started by spreading over this a second layer, acidulated with nitric or hydrochloric acid.

§ 41. Boron

a. Precipitation as potassium fluoborate. Limit: 0.2 μgr. of B.[1]

b. Sublimation with ammonium fluosilicate. Limit: 0.4 μgr. of B.

a. Dissolve in hydrofluoric acid or in hydrochloric acid

[1] Behrens, Mikr. Meth. p. 38.

and ammonium fluoride and add potassium chloride or nitre. Dilute solutions must be concentrated. They will then yield better crystals than the thin, pale rhombs, hexagons, and octagons (orthorhombic), which are produced by rapid precipitation. Recrystallisation from hot water yields prismatic crystals. By substitution of Rb or Cs for potassium little is gained. If silicon be present in considerable quantity, a great part of it may be volatilised by heating on the water bath. The rest is precipitated with barium acetate as $BaSiF_6$.

FIG. 51.—Fluoborate of potassium, × 130.

b. Stolba has found that boron may be volatilised by heating its compounds with ammonium fluosilicate. The sample is ground up with half its volume of the fluosilicate and subjected to sublimation in a small test tube. The sublimate is dissolved in very dilute hydrochloric acid, silicon is precipitated as $BaSiF_6$; from the mother liquor nitre will precipitate KBF_4.

Boric acid has an injurious influence on several reactions, of the same kind as aluminium and iron. See §§ 19-21 (Barium, Strontium, and Calcium).

§ 42. Aluminium

a. Precipitation as cæsium alum. Limit: 0.35 μgr. of Al.[1]

b. Precipitation with ammonium fluoride. Limit: 0.3 μgr. of Al.

a. Evaporate with a small drop of sulphuric acid, dis-

[1] Behrens, l.c. p. 30.

solve the residue in water, and put a grain of cæsium chloride into the drop near to its border. The concentration is an essential point. If more than 1 per cent of aluminium sulphate be present, rectangular dendrites are seen to sprout from the cæsium chloride. A small drop of water must then be added where the reagent has been put in. With less than 0.2 per cent of aluminium sulphate it is difficult to obtain good crystals. Concentration by heat always necessitates long waiting. Spontaneous evaporation is yet more tedious, but it gives very fine crystals. A slight excess of sulphuric acid is favourable, a great excess of strong acids must be made innocuous by adding an acetate. Cæsium alum forms colourless octahedra (40 to 90 μ) of great beauty. Streng prescribes cæsium sulphate instead of the chloride. Its action is more rapid, and no excess of sulphuric acid is necessary. Acid potassium sulphate, also proposed by Streng, produces beautiful crystals, but the test loses its delicacy. Ferric compounds will not interfere with this test, ferric alum being difficult to crystallise. Great quantities of alkali salts are far more injurious, their crystals interfering with the crystallisation of the octahedra of cæsium alum. Testing for aluminium ought therefore not to be deferred. Otherwise, precipitation with caustic ammonia, or boiling with a dilute solution of ammonium acetate, must be employed for separating aluminium hydroxide from the alkali salts. A small blotting-pad is made by folding a strip of filtering-paper three or four times; on this small drops of the precipitated liquid are put from a pipette on the same spot. Finally, the bit of paper, on which the precipitate

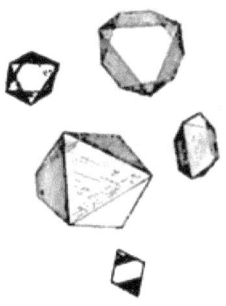

FIG. 52.—Cæsium alum, ×130.

has been collected, is cut out; it is washed on another pad with a drop of water, and heated on a slide with dilute hydrochloric acid. The solution is run aside, the paper is washed with a drop of water, and the operation is completed by concentrating and testing.

b. Ammonium fluoride, added in excess to solutions of aluminium, precipitates pale, well-developed octahedra of the compound $AlF_3 \cdot 3NH_4F$. In dilute solutions they are slowly developed along the periphery of the drop, growing to a size of 60 μ. They dissolve easily in nitric acid; by ammonium acetate they are reprecipitated. The fluorides of potassium, rubidium, and cæsium have a more energetic action, but the crystals are smaller. Sodium and lithium fluorides will produce pale, granular precipitates. Precipitation with ammonium fluoride will be found serviceable for very dilute solutions, in which it enables us to detect aluminium in a shorter time than is possible with cæsium chloride. Sodium and ferric compounds must, however, be absent. Preference will generally be given to test a.

FIG. 53.—Fluoaluminate of ammonium, the short rods with excess of Al, ×130.

For staining aluminium hydroxide, Congo red may be employed. Its aqueous solution produces a blood-red colour at ordinary temperature more speedily and with greater intensity on gelatinous hydroxide than on the same substance after it has been dried.

§ 43. Iron

a. Precipitation with potassium ferrocyanide. Limit: 0.07 μgr. of Fe.

b. Precipitation with ammonium fluoride. Limit: 0.2 μgr. of Fe.

c. Precipitation with barium acetate and oxalic acid. Limit : 0.1 μgr. of Fe.

a. The blue flakes and films of Prussian blue are distinctly seen under powers not exceeding 200. The limit of dilution is 1 part of ferric chloride in 5000 parts of water. Great quantities of strong acids must be avoided, as they might cause production of Prussian blue by decomposing the reagent.

b. Ammonium fluoride precipitates ferric solutions in the same manner as solutions of aluminium. The crystals are less pale, otherwise they resemble those described in § 42, *b.* Sodium and lithium fluorides precipitate globules of 6 to 12 μ, silver fluoride precipitates a light yellow powder. Precipitation with ammonium fluoride is of greater value in the case of iron than of aluminium, as the presence of iron in the crystals may easily be established with caustic ammonia. For this purpose the crystals are rapidly washed; they are transferred in a drop of water to an ordinary slide, freed from water with a roll of filtering-paper, and covered with a drop of ammonia. Crystals containing iron are immediately stained yellow or brown without losing their shape. Transformation to Prussian blue will not answer.

FIG. 54.—Double oxalate of barium and iron, ×60.

c. Ferric chloride modifies the crystallisation of barium oxalate in a most remarkable manner. Instead of short colourless rods there will sprout from the periphery of the drop tufts of light brown curling hairs,[1] growing towards the centre, and reaching a length of 300 μ. A great excess of reagents is injurious. Aluminium compounds will, under the same circumstances, produce similar colourless growths. Strontium oxalate is modified to rectangular prisms, calcium oxalate does not undergo any perceptible change.

Ferrous compounds are precipitated by oxalic acid, even from strongly acidulated solutions.[2] This reaction is slow, and the small yellowish prisms of ferrous oxalate are not characteristic.

The double *phosphate*, precipitated by sodium phosphate from ammoniacal solutions of ferrous compounds, agrees with ammonium-manganous phosphate. It is not fitted for microchemical use, because the reduction of ferric compounds in small drops is beset with difficulties.

§ 44. Chromium

a. Precipitation as chromate of silver. Limit: 0.025 μgr. of Cr.

b. Precipitation as chromate of lead. Limit: 0.02 μgr. of Cr.

a. For the properties of silver chromate, see § 7, *b*. Where chlorides do not stand in the way, precipitation of silver chromate will be preferred, although the delicacy is inferior to that afforded by the chromate of lead. Its large and characteristic crystals are easily distinguished from those of the vanadate and arsenate of silver. The delicacy may be increased by making use of the isomorphism

---

[1] Platinum tetrachloride yields with barium oxalate hair-like growths resembling in all respects those described above.

[2] Haushofer, l.c. p. 49.

of the chromate and sulphate of silver. If the sulphate predominates, a great quantity of nitric acid is not able to bring about the crystallisation of the bichromate $Ag_2Cr_2O_7$. The normal chromate Ag_2CrO_4 is invariably formed entering into the orthorhombic crystals of Ag_2SO_4 and staining them from light yellow to fiery red, according to the proportion of chromium to sulphur. This modification of the test is especially useful if the solution be charged with alkali salts.

b. By precipitation at ordinary temperature no crystals are produced. Bourgeois has shown that characteristic crystals are formed if chromate of lead is precipitated from a hot solution, to which a moderate quantity of nitric acid has been added. Lenticular crystals and short, slender prisms ($20\ \mu$) are formed. The latter are characteristic, even if the yellow colour is not distinctly seen.

FIG. 55.—Chromate of lead, ×60.

Chromic compounds are converted into chromates: (1) by fusion with sodium carbonate and a little nitre; (2) by slightly heating an ammoniacal solution with hydrogen peroxide; (3) by boiling their solution in nitric acid with potassium chlorate till the colour of the solution has become pure yellow or orange. The first method is the safest and suited for all chromic compounds; for microchemical use it has the inconvenience that a great quantity of alkali salt is introduced into the solution. Here the compound crystals of sulphate and chromate of silver are very useful. The second method is excellent for testing with silver nitrate, provided that chlorides are absent. The third method relies exclusively on test *b.* It is particularly fitted for the examination of chromium alloys. See Ferrochrome, in Part II., § 121, b.

§ 45. Vanadium

a. Precipitation as ammonium vanadate. Limit: 0.3 μgr. of V.[1]

b. Precipitation as silver vanadate. Limit: 0.07 μgr. of V.

c. Precipitation as thallous chlorovanadate. Limit: 0.07 μgr. of V.

a. From alkaline solutions of vanadic acid, ammonium chloride will precipitate small lenticular crystals (15 to 20 μ) of ammonium metavanadate, provided that a sufficient quantity of the reagent be added to form a saturated solution of ammonium chloride, in which ammonium vanadate is almost insoluble. The little crystals are colour-

Fig. 56.—Metavanadate of ammonium, × 130.

less; their strong polarisation is extinguished in the direction of their length.

Fig. 57.—Pyrovanadate of silver, × 130.

b. Colourless solutions of metavanadates are coloured yellow by acids. From such acidulated solutions silver nitrate precipitates a yellow powder and very small prisms. If silver nitrate is added to a hot solution, strongly acidulated with acetic acid, orange-coloured rods (10 μ) are precipitated, often grouped to fans and stars (20 to 30 μ). These crystals cannot be confounded with silver bichromate.

[1] Streng, Anl. z. Best. d. Miner. p. 89.

c. Thallous nitrate precipitates a white powder from solutions of metavanadates, insoluble in water, soluble in dilute solutions of ammonium chloride. From such a solution, on cooling, small crystals (10 to 15 μ) separate out, resembling crystals of lead sulphate. They dissolve readily in dilute nitric acid. From this solution orange-coloured rhombs and dendrites of thallous pyrovanadate [1] separate out, which may be confounded with thallous chromate. If an excess of ammonia be added to the hot solution, light yellow hexagonal rosettes (20 to 30 μ) will crystallise. This reaction is rather slow, but it is characteristic and of sufficient delicacy. The characteristic button-like crystals are not formed if no chloride is present. They are probably crystals of a chlorovanadate.

§ 46. Niobium

a. Precipitation as sodium niobate. Limit: 0.6 μgr. of Nb.[2]

a. Sodium niobate ($NaNbO_3 + 3H_2O$) forms colourless rods (probably orthorhombic) of about 30 μ, often grouped to six-spoked stars which may develop into six-sided plates, resembling those of sodium tantalate. Rapid crystallisation yields needles like those of gypsum. If the solution contains much caustic alkali, stars and six-sided plates will appear after a time. Slow crystallisation from solutions containing a small quantity of caustic alkali yields prismatic

FIG. 58.—Niobate of sodium, ×130.

[1] Haushofer, l.c. p. 135.   [2] Ibid. p. 104.

crystals. If insoluble compounds of niobium are fused with sodium carbonate or with caustic soda, a small quantity of water will dissolve very little niobate, because sodium niobate is almost insoluble in strong solutions of soda. A second treatment with hot water yields a solution from which crystals may be obtained by concentration or by precipitation with caustic soda. Niobic acid, prepared by fusing with acid potassium sulphate and lixiviation with water, is dissolved by heating with a weak solution of caustic soda. Fusion of columbite with caustic potash yields a mass easily dissolving in water. From this solution sodium salts and caustic soda will precipitate crystals of sodium niobate. For complete precipitation much time must be allowed. Sodium niobate dissolves in 200 parts of water at ordinary temperature, in 75 parts of boiling water. It dissolves freely in hydrochloric acid if ammonium fluoride is added, and this solution is not precipitated by potassium compounds.

Hydrated niobic acid is stained orange or brick-red by extract of galls if a little hydrochloric acid be added. This reaction is of little value for microscopical observation; it is of no use whatever for tracing niobic acid, mixed with a considerable quantity of tantalic acid.

§ 47. Tantalum

a. Precipitation as fluotantalate of potassium. Limit: 6 μgr. of Ta.

b. Precipitation as tantalate of sodium. Limit: 1.2 μgr. of Ta.[1]

a. From solutions of tantalic acid in hydrofluoric acid

[1] Haushofer, l.c. p. 104.

or in hydrochloric acid, mixed with ammonium fluoride, potassium salts precipitate the compound K_2TaF_7, crystal-

Fig. 59.—Fluotantalate of potassium, ×130.

lising in long slender prisms (50 to 200 μ). It dissolves at ordinary temperature in 160 parts of water, while potassium fluoxyniobate formed under the same circumstances dissolves in 13 parts of water. However sparingly soluble, the fluotantalate is rather slow to crystallise. For complete crystallisation at least ten minutes must be allowed. With cæsium the difference of solubilities is much greater, but unfortunately the crystals of the compound Cs_2TaF_7 will scarcely reach a length of 1.5 μ.

b. Tantalic acid is more difficult to dissolve in solutions of caustic soda than niobic acid. If both are present in an alkaline solution, sodium tantalate is precipitated first if caustic soda be added. Sodium tantalate ($Na_8Ta_6O_{19} + 25H_2O$) dissolves at ordinary temperature in 490 parts of water, at a boiling heat it requires 162 parts of water. It is advisable to put it into boiling water if it is to be dissolved. If cold water is used much more caustic alkali is needed than for sodium niobate, to prevent precipitation of acid tantalates. The crystals of sodium tantalate (30 to 50 μ) have usually the form of the crystals of sodium fluosilicate, but they are colourless. It must, however, be borne in mind that sodium niobate may also assume the form of hexagonal plates and rosettes, differing from those of the tantalate in nothing save their greater size. Further, if sodium tantalate is slowly precipitated with sodium chloride from a solution of tantalic acid in caustic potash, it exhibits the prismatic form of the niobate. For these reasons test a, though less delicate and less expeditious, is the principal one. The prismatic crystals of potassium fluotantalate are at ordinary temperature slowly corroded by a solution of caustic soda; a slight heating will complete the decomposition, and on cooling, crystals of sodium tantalate will separate out.

Hydrated tantalic acid is stained light yellow by extract of galls and hydrochloric acid. This test is of no value for microchemical observation.

§ 48. Bismuth

a. Precipitation with acid potassium oxalate. Limit: 0.3 μgr. of Bi.

b. Precipitation with potassium iodide and rubidium chloride. Limit: 0.13 μgr. of Bi.

c. Precipitation with acid potassium sulphate. Limit: 0.3 μgr. of Bi.

d. Decomposition of the chloride or iodide with water. Limit: 0.4 μgr. of Bi.

a. Oxalic acid produces in solutions of bismuth a dense powdery precipitate composed of colourless tiny needles. Acid potassium oxalate produces the same precipitate more speedily and completely. By heating the precipitate is dissolved; on cooling, a double oxalate of potassium and bismuth separates out in highly refractive and strongly polarising crystals of octahedral shape (6 to 15 μ), resembling crystals of calcium oxalate. This test succeeds in solutions containing 0.1 per cent of bismuth nitrate.

FIG. 60.—Double oxalate of potassium and bismuth, ×300.

b. Solutions of bismuth compounds in hydrochloric acid

yield with rubidium chloride colourless six-sided plates (probably orthorhombic), measuring from 200 to 300 μ. They are very thin, sharply outlined, and strongly polarising. If tin be present, bismuth is precipitated in the second place; its crystals will then take the form of oblong hexagons. Rubidium chloride affords sufficient delicacy for solutions containing 0.2 per cent of bismuth nitrate. For very dilute solutions cæsium chloride must be employed, yielding smaller crystals of the same form. This test is convenient if tin must be traced also. It is very beautiful if a little potassium iodide be added. The iodide acts more slowly than the chloride, but surely and in a characteristic manner. The drop will yield to the last, rods, rhombs, and hexagons of a blood-red colour, smaller than the crystals of the double chloride.

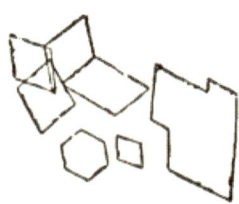

FIG. 61.—Double chloride of bismuth and rubidium, ×130.

 c. See § 1, *c.* This reaction requires solutions containing at least 0.2 per cent of bismuth nitrate. On the other hand, it is more decisive than test *a.*

 d. Precipitation of bismuth with arsenic acid and decomposition of bismuth nitrate with water[1] are not to be recommended. I have found them tedious and wanting in neatness. Decomposition of the chloride with water will take place in the presence of a considerable quantity of hydrochloric acid, and decomposition of the iodide will give yet better results. Potassium iodide produces in acid solutions a dark cloud, dissolving in a slight excess

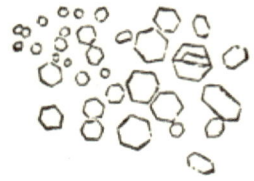

FIG. 62.—Double sulphate of bismuth and potassium, ×130.

[1] Haushofer, l.c. p. 138.

of the reagent with a dark yellow colour. This solution is decomposed by water. A dark brown precipitate is formed, turning to brownish-red if heated in the liquid. If antimony be present a dark red precipitate is formed. In the presence of tin it is a reddish-orange. Too much potassium iodide will keep a considerable quantity of bismuth dissolved. In this case a grain of lead acetate will produce a dark red precipitate, appearing in very dilute and strongly acidulated solutions in the course of evaporation. Water is then added to complete the precipitation.

§ 49. Antimony

a. Precipitation with cæsium chloride. Limit : 0.16 µgr. of Sb.

b. Precipitation with oxalic acid. Limit : 1.0 µgr. of Sb.

c. Precipitation as double tartrate of barium and antimonyl. Limit : 1.0 µgr. of Sb.[1]

d. Precipitation as sodium antimonate. Limit : 0.5 µgr. of Sb.[2]

a. It is necessary to proceed as indicated for bismuth, § 48, *b.* If a considerable quantity of bismuth be present, the double chloride of bismuth is the first to crystallise. The crystals of chlorostibite attain 300 µ with rubidium chloride, 80 µ with cæsium chloride. They resemble the crystals described in § 48, *b*, only they are paler. Addition of potassium iodide will carry the limit to 0.08 µgr., and will produce crystals of an orange colour, appearing (from

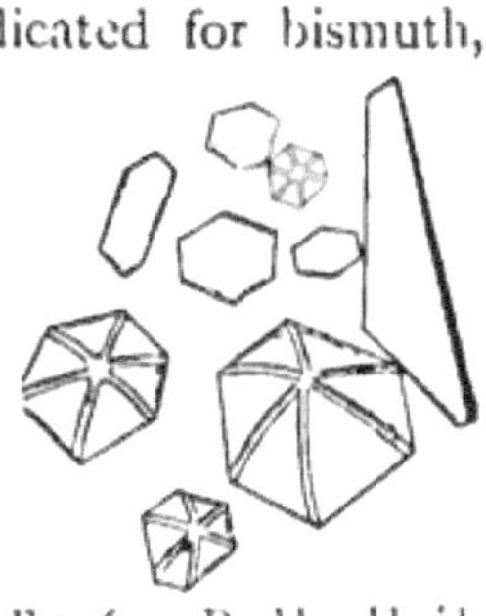

FIG. 63. — Double chloride of cæsium and antimony, ×130.

---

[1] Streng, N. Jahrb., 1885, i. p. 53.   [2] Haushofer, l. c. p. 14.

acid solutions) after the crystals of the bismuth compound.

b. With oxalic acid or with acid potassium oxalate brushes and fans of thin, thread-like crystals of antimonious oxalate (oxalate of antimonyl, $C_2O_4 2SbOH + 2H_2O$) are produced. This compound dissolves in hydrochloric acid; by water it is reprecipitated, apparently unaltered. This convenient test may be taken as characteristic, if tin, lead, and bismuth are absent; by these metals it may be masked.

c. The property of tartaric acid of dissolving antimonious hydroxide and antimonious oxychloride may be turned to account for separating antimony from bismuth. The solution is tested with cæsium chloride (§ 49, *a*). Antimonious oxychloride may also be dissolved by boiling with water and barium tartrate, producing the double tartrate described in § 19, *d.* Streng recommends the precipitation of the antimony beforehand as sulphide, and the preparation from this of the oxychloride needed for the test. If this plan be followed, the test will succeed without difficulty, and may be considered as conclusive.

FIG. 64.—Antimonate of sodium, × 130.

d. The sample must be fused with five times its volume of nitre, raising the heat to a bright red, and keeping it up till half of the nitre has been volatilised. The assay must be washed with cold water, boiled with a weak solution of caustic potash, and then a grain of sodium chloride must be added to the clear liquid. The crystals of sodium pyroantimonate ($Na_2H_2Sb_2O_7 + 6H_2O$) are lenticular (20 to 50 μ) if rapidly formed; from dilute solutions larger prisms will slowly separate out, strongly polarising, extinguishing between crossed nicols parallel to the edges of the prisms. Three crystals are often seen crossing each other at right angles. If the crystallisation is too slow, it may be accelerated by a small drop of alcohol. This test is decisive if *niobium* is not present.

§ 50 and § 51. Arsenic

(1) *Arsenious Oxide*

a. Precipitation from alkaline solutions with nitric acid. Limit: 0.14 μgr. of As = 0.2 μgr. As_2O_3.

a. Evaporation of ammoniacal solutions will not answer for microchemical examination. Sublimation and treatment with hot hydrochloric acid have likewise not given a good result. Well-developed octahedra (6 to 10 μ) have been obtained by acidulating a solution of arsenious oxide in caustic potash (excess of solvent is to be avoided) with nitric acid, and concentrating at a gentle heat. The crystals are chiefly formed at the periphery of the drop. By sulphuretted hydrogen they are coloured yellow without losing their form and their limpidity.

Silver arsenite is precipitated from mixed solutions of silver nitrate and arsenious oxide if a trace of caustic ammonia is added. It forms pointed rhombs and needles (30 μ) of the colour of sulphur. This compound is dissolved by a slight excess of acid, and also by ammonia.

(2) *Arsenic Acid*

a. Precipitation as ammonium calcium arsenate. Limit: 0.035 μgr. of As.

b. Precipitation as ammonium arsenomolybdate. Limit: 0.22 μgr. of As.

a. Oxidation of arsenious oxide, of arsenides, etc., is best effected by heating gently with hydrochloric acid and potassium chlorate. A great excess of hydrochloric acid is injurious if precipitation with a calcium compound is intended, as ammonium chloride is a solvent for ammonium-calcium arsenate. This precaution is unnecessary if ammonium magnesium arsenate is to be precipitated. The solution is gently heated with an excess of ammonia, and precipitated with a grain of calcium acetate.

Fig. 65.—Double arsenate of calcium and ammonium, × 130.

The crystals are of the same type as those of ammonium magnesium phosphate, only they are longer. From highly diluted solutions rod-like crystals of 15 μ are obtained, showing only traces of hemimorphism. This test is even as delicate as that with magnesium compounds, and it has the advantage that it excludes phosphoric acid. One weak point has already been noticed, viz. the dissolving action of ammonium chloride; another, yet to be noted, is the tendency of calcium to form insoluble compounds with a great many acids, which renders this test impracticable in the presence of carbonates, and of oxalates, and for determining the presence of arsenic acid in the precipitate produced by means of ammonium molybdate (§ 51, *b*). In all such cases *zinc* must take the place of calcium. The test is then even as delicate and decisive. The crystals of ammonium zinc arsenate show the same forms as those of ammonium calcium arsenate.

b. At ordinary temperatures precipitation of ammonium arsenomolybdate is exceedingly slow. The precipitate, formed at a temperature of about 40°, offers the same aspect as potassium phosphomolybdate (§ 1, *a*). In case of need, ammonium molybdate may be used to precipitate phosphoric acid from an acid solution at ordinary temperature, and later, at a gentle heat, arsenic acid. The test *a* is, however, safer and more delicate. Further, a precipitate of ammonium arsenomolybdate may contain vanadium and bismuth, which will form complex molybdates of the same appearance and under the same circumstances. If much ammonium chloride is present, even silicomolybdate may separate out.

§ 52. Phosphorus

a. Precipitation as ammonium magnesium phosphate. Limit: 0.008 μgr. of P.

b. Precipitation as ammonium phosphomolybdate. Limit: 0.015 μgr. of P.

a. This test is of particular value for the examination of compounds that have been fused with alkali. If a little potassium cyanide has been added before the fusion, all chance of confounding phosphorus and arsenic is excluded by the volatilisation of the latter element. In solutions, arsenic acid is reduced to arsenious oxide by sulphur dioxide. For the properties of ammonium magnesium phosphate see § 8, *a*. The operation of testing undergoes a little modification if fusion with alkaline carbonates has been employed. Close to the drop of the aqueous solution from the fused mass a drop of water is placed, in which some ammonium chloride and a little magnesium acetate are dissolved. The slide is then heated, and the two drops are made to flow together by putting a drop of ammonia between them.

b. A very convenient test for acid solutions of insoluble phosphates, and of the greatest value in the examination of ground specimens of rocks. The reagent is prepared by dissolving ammonium molybdate in warm dilute nitric acid. The solution must be clear, and must not deposit yellow grains during spontaneous evaporation. Precipitation of ammonium phosphomolybdate commences at ordinary temperature. It is indeed accelerated by heat, but the operator risks precipitating at the same time arseno-molybdate (prevented by treating the original acid solution with sulphur dioxide), and even silicomolybdate, if much ammonium chloride is present. The aspect of ammonium phosphomolybdate agrees in all points with that of potassium phosphomolybdate (§ 1, *b*). For tracing phosphorite and apatite on specimens of rock prepared for the microscope, the reagent must be spread in a thin layer, and it must be prepared as concentrated and as acid as possible, with a view to localise the reaction and to dispense with heating. Heating would interfere with the localisation, and there would be a chance of producing silicomolybdate, as even orthoclase is attacked by hot, strong acids.

§ 53. Nitrogen

a. Tracing of nitrites with potassium iodide and starch. Limit: 0.25 μgr. of HNO_3.

b. Precipitation of ammonia with platinum tetrachloride. Limit: 0.1 μgr. of NH_3.[1]

c. Tracing of cyanogen with iron compounds. Limit: 0.07 μgr. of Cy.

a. Nitrates may be reduced to nitrites in the wet way by means of metallic magnesium. The operation may be conducted on a slide. The neutral solution is heated for two or three minutes with a small quantity of magnesium powder, a drop of water being added from time to time. Reduction in the dry way is far more energetic. A piece of iron or nickel wire is coated with lead by dipping it first in a solution of the double chloride of ammonium and zinc, and afterwards in the molten metal. On the coated point, bent to a small hook, the nitrate is kept fused till the lead shows a yellow crust. A dull red heat is suited for this operation. The fused mass is dissolved in water; it is acidulated with sulphuric acid, and tested with a trace of potassium iodide and a few grains of starch (§ 62, *f*). If iron or manganese be present, great caution is necessary, in order to avoid errors. Test *b* is then to be preferred.

b. Distillation of ammonia may be avoided by employing precipitation of mercuric chloride, or by inverting test § 8, *a*. Chloride of mercurammonium is a white powder. This is an objection to the reaction, otherwise recommended by its great delicacy (limit: 0.05 μgr. of NH_3). The same objection may be made to the use of Nessler's reagent, which produces yellow flakes. On the contrary, the crystals of ammonium magnesium phosphate may always be recognised if high powers be employed. In the present case a little magnesium acetate is added to the sample, sodium phosphate and sodium bicarbonate are

[1] Behrens, Mikr. Meth. p. 29.

dissolved in a drop of water placed close to the sample. After a slight heating the drops are made to flow together. If calcium be present caustic soda must be employed. In this case, however, it will be difficult to avoid flakes of magnesium hydroxide. The limit may be carried to 0.05 μgr. of NH_3.

Distillation of ammonia on the slide is managed in the following manner:—The sample is evaporated with a little dilute hydrochloric acid. The residue is encircled with a small ring bent out of iron or nickel wire of about 1.5 mm. thickness, which must support the cover. A piece of filtering-paper is cut a little smaller than the ring, and on a cover-glass a small drop of hydrochloric acid is spread out. Then a drop of a strong solution of caustic soda is put on the residue within the ring, the paper is laid on, which absorbs the liquid and prevents spurting, the cover-glass is put upon the ring, and on the cover-glass a drop of water or alcohol to cool it. Heat is applied till a fine dew is seen to settle round the small drop of hydrochloric acid. If a platinum spoon be used instead of the slide, a small tuft of asbestos is put upon the liquid. In a spoon this method may be applied to nitrates by adding fine zinc filings with the caustic soda. The result may rival that of method a. After distillation all the methods described in § 1, a to c, may be employed for testing. Preference will generally be given to an acidulated solution of platinum tetrachloride. The crystals of ammonium chloroplatinate agree in all respects with those of the potassium compound (§ 1, a).

c. The majority of cyanides may be decomposed by heating with a strong solution of caustic potash. The strongly alkaline solution is afterwards heated with a mixture of ferric and ferrous chlorides; finally a drop of dilute hydrochloric acid is added without disturbing the precipitate. The particles of Prussian blue will gradually come into view as the ferric hydroxide is dissolved. Cyanide of mercury must be reduced on bright iron in a strong solution of caustic potash. Decomposition is accelerated by contact with platinum.

§ 54. Sulphur

a. Precipitation as sulphate of calcium. Limit: 0.2 μgr. of S.

b. Precipitation as cæsium alum. Limit: 0.12 μgr. of S.

c. Precipitation as sulphate of lead. Limit: 0.006 μgr. of S.

a. Sulphides must be oxidised by fusion with sodium carbonate and nitre. Insoluble sulphates are fused with sodium carbonate. The fused mass is treated with water, the clear solution is acidulated with acetic acid and tested with calcium acetate (see § 21, *a*). The limit of instantaneous reaction is found when a dilution of 1 part of sodium sulphate in 400 parts of water is reached. In very dilute solutions great quantities of alkali salts are injurious.

b. Tracing sulphuric acid as cæsium alum by means of cæsium chloride and aluminium nitrate gives better results in dilute solutions, overcharged with alkali salts, than method *a*. This advantage is not due to feeble solubility, the limit of dilution for instantaneous reaction being the same for both methods; it is rather a consequence of the greater size and characteristic form of the crystals of cæsium alum. For detail, see § 42, *a*.

c. For very dilute and impure solutions precipitation

with acetate of lead may be employed. Here the limit of instantaneous reaction is found when a dilution of 1 part of sodium sulphate in 4000 parts of water is reached. This test is rather slow, and requires a power of 300 and a practised observer. It is, however, of great value for its delicacy, and as affording the possibility of working with solutions charged with great quantities of salts and acid.

§ 55. Selenium

a. Reduction of selenium dioxide with magnesium. Limit : 0.1 μgr. of Se.

b. Reduction with stannous chloride. Limit : 0.5 μgr. of Se.

c. Precipitation with potassium iodide. Limit : 1 μgr. of Se.

a. Reduction of selenium dioxide with a solution of sulphur dioxide [1] in water is very tedious, and introduces a great quantity of water into the solution. The limit given by the author of this method is indeed very low—20 μgr. of Se. If zinc or magnesium be used as reducing agents in solutions of selenium dioxide, acidulated with acetic acid, the delicacy is raised at least a hundredfold. This high degree of delicacy is rather unexpected, considering that the product of the reaction is an element, instead of a compound, of great molecular volume, as in the case of double phosphates or triple nitrites. It is principally owing to the fact that selenium is deposited as a coherent crust or shell on the metallic particles. After some minutes the remaining metal may be dissolved in hydrochloric acid, when red shells, which are the hollow,

[1] Streng, N. Jahrb., 1886, i. p. 51.

transparent casts in selenium of the metallic particles, will remain. If less than 0.02 per cent of selenium be present only red flakes are produced.

b. Stannous chloride reduces selenium from solutions strongly acidulated with hydrochloric acid. Scarcely a quarter of the delicacy of test a is reached. Nevertheless, this reagent is useful on account of its rapid action and of the characteristic red colour of the precipitate.

c. For the same reason, precipitation with potassium iodide deserves notice. In solutions of selenates no precipitation takes place, unless they have been previously heated with hydrochloric acid. Neutral solutions of selenium dioxide, and such as are acidulated with acetic acid, are also not precipitated by potassium iodide. If hydrochloric acid be added, the liquid turns yellow, as if bismuth were present; a smoky film spreads towards the centre, and a brownish-red powder of SeI_4 is thrown down. It is more volatile than selenium. Sublimates of both cannot be distinguished by their appearance. In a solution of potassium iodide or sodium iodide it dissolves freely, forming red and orange rods and plates, looking like iodide of antimony. They are decomposed by water, leaving powdery selenium tetraiodide.

Soluble selenates will precipitate solutions of calcium in the same manner as sulphates. Calcium selenate is very slowly decomposed by hot hydrochloric acid.

§ 56. Tellurium

a. Reduction of the dioxide with magnesium. Limit: 6 μgr. of Te.

b. Precipitation as chlorotellurite of cæsium. Limit: 0.3 μgr. of Te.

c. Precipitation with potassium iodide.   Limit : 0.6 μgr. of Te.

a. Reduction of tellurium dioxide with sulphur dioxide[1] has the same defects as analogous reduction of selenium dioxide.   With metallic zinc or magnesium small scales and films of tellurium are produced, looking dark grayish-brown in transmitted light.   A considerable quantity of selenium may be masked in these scales, not betrayed by sublimation.   On the other hand, tellurium itself is masked by arsenic.

FIG. 66.—Chlorotellurite of cæsium, ×130.

b. Cæsium chloride precipitates from solutions of tellurium dioxide in hydrochloric acid yellow octahedra (10 to 30 μ) of the compound Cs_2TeCl_6.   They are larger and less refractive than crystals of cæsium chloroplatinate ; they are besides decomposed by water.   From chlorostannate they are easily distinguished by means of potassium iodide : chlorostannate is stained yellow, chlorotellurite is blackened.

c. With potassium iodide no reaction is perceived in alkaline solutions of tellurium dioxide.   If hydrochloric acid is added, the liquid becomes dark yellow, a brown metallic film is formed, growing black when dried.   With more potassium iodide a brown liquid is produced, from which dark rhombs, hexagonal grains, and rods (10 to 20 μ) come down.   In reflected light the colour of these crystals (probably of the compound TeI_4) is a reddish-brown.   By bismuth this reaction may be made useless, tellurium being masked.   If tellurium is associated with selenium, potassium iodide will precipitate first TeI_4, later, after a slight heating,

---

[1] Haushofer, l.c. p. 124.

a red ring of SeI$_4$ is formed. After evaporation, the precipitate may be washed, dried, and subjected to sublimation, when SeI$_4$ will be sublimed a long time before the tellurium compound is volatilised.

§ 57. Molybdenum

a. Precipitation as phosphomolybdate. Limit: 0.1 μgr. of Mo.[1]

b. Precipitation as thallous molybdate. Limit: 0.033 μgr. of Mo.

a. For the properties of ammonium phosphomolybdate see § 52, *b*; for potassium phosphomolybdate see § 1, *b*.

A difficulty arises from the circumstance that phosphomolybdates dissolve in solutions of alkaline phosphates. Great caution is therefore necessary not to add more than a trace of sodium phosphate. Besides the two phosphomolybdates, named above, thallous phosphomolybdate demands notice. Its grains are smaller than those of the ammonium compound, dark yellow, clinging to the glass, and thus very suitable for separating and accumulating molybdic acid. Similar yellow precipitates are formed by heating thallous molybdate with nitric acid. They seem to be thallic-molybdates, analogous to silicomolybdates; either potassium thallic-molybdate or thallous thallic-molybdate.

b. From solutions of molybdic acid in a slight excess of caustic alkali a grain of thallous nitrate will precipitate glittering six-sided plates (30 to 60 μ), colourless or iridescent. Sometimes six-spoked stars are produced. This compound, thallous molybdate (Tl$_2$MoO$_4$),[2] dissolves in hot

---

[1] Haushofer, l.c. p. 97.
[2] Oettinger, Zeitschr. f. Chem. u. Pharm., 1864, p. 440.

water, recrystallising slowly, in confused crusts. By nitric acid it is rapidly decomposed and dissolved. Heating of the acid solution precipitates the yellow powder described as a thallic-molybdate in *a*. Addition of a trace of sodium phosphate precipitates thallous phosphomolybdate. On the other hand, thallous phosphomolybdate is decomposed by caustic alkali, iridescent scales of thallous molybdate being produced. The limit of instantaneous crystallisation of thallous molybdate is found when a dilution of 1 part of molybdic acid in 5000 parts of water is reached.

Fig. 67.—Thallous molybdate, ×130.

Lead acetate may be employed instead of thallous nitrate; it produces scales of less striking appearance. From acid solutions it precipitates square plates (tetragonal system), which might be utilised if better tests were not available.

The same may be said of the crimson tint produced in acid solutions of molybdic acid by soluble thiocyanates and metallic zinc or magnesium. It is very well perceived under a low power, and may be of service for distinguishing molybdic from tungstic acid.

§ 58. Tungsten

a. Precipitation of tungstic acid with stronger acids. Limit: 1.6 µgr. of W.

b. Precipitation as phosphotungstate. Limit: 0.12 µgr. of W.

c. Precipitation as thallous tungstate. Limit: 0.08 µgr. of W.

a. The surest way to produce the characteristic yellow

colour of tungstic acid is to heat the white hydrate with strong hydrochloric acid. The heat must be raised to ebullition, and the hydrated tungstic acid must be in a compact state. Loose flakes or fine powder will take a faint tint only. For this reason the experiment is limited to at least 10 μgr. of tungstic acid. No change of structure is perceived. The yellow acid is not crystalline; it is composed of flakes and small grains. This is to be borne in mind to prevent confusion with phosphomolybdates and similar compounds. If the experiment does not succeed, the different solubility of tungstic and molybdic acids may be turned to account. Evaporate with nitric acid, and extract molybdic acid by heating the residue with dilute nitric acid, which leaves tungstic acid behind. Small rhombs are to be attributed to acid tungstate; if they are perceived, evaporation and extraction must be repeated.

b. The crystals of potassium and ammonium phosphotungstate have the same form and size as those of the phosphomolybdates, but they are colourless. From mixed solutions compound crystals are produced, more or less yellow, according to the quantity of molybdenum.

c. See § 57, *b.* The crystals of thallous tungstate are about thrice as large as those of the molybdate; they may reach 400 μ. By nitric acid and a trace of sodium phosphate they are transformed to granular phosphotungstate.

Barium and calcium tungstate, recommended by Haushofer,[1] form very small, insignificant crystals. Ammonium tungstate[2] crystallises fairly well, and will leave on calcination a characteristic bluish residue. But

Fig. 68.—Thallous tungstate, ×60.

---

[1] Haushofer, l.c. pp. 144, 145.   [2] Ibid. p. 146.

the considerable quantity of tungstic acid required for this experiment—at least 100 μgr.—forbids its use for a microchemical test.

§ 59. Uranium

a. Precipitation with sodium acetate. Limit: 0.6 μgr. of U.[1]

b. Precipitation from ammoniacal solutions with thallous nitrate. Limit: 0.1 μgr. of U.

a. The limit given above for the test with sodium acetate is reached only if evaporation is pushed to the extreme limit. Instantaneous reaction is limited by a dilution of about 1 part of uranyl acetate in 150 parts of water. For more particulars, see § 2, *a*.

b. Acid solutions of uranyl are precipitated by ammonia. The precipitate dissolves readily in a strong solution of ammonium carbonate. In a solution of this kind a grain of thallous nitrate is covered with short, pointed crystals. At a distance, clear, well-developed rhombs (30 to 70 μ) are formed, pale yellow, vividly polarising. Their polarisation is extinguished in the direction of their length. This reaction goes a long way farther than *a*; it can be perceived, with all necessary distinctness, at a dilution of 1 : 5000. Its result may be checked by treating the crystals with a solution of potassium ferrocyanide in acetic acid. They are dissolved, reddish grains coming in their place. If the presence of copper is suspected, sodium carbonate may be employed instead of ammonium car-

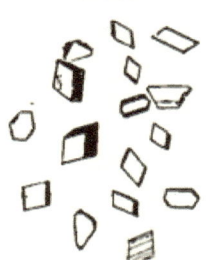

Fig. 69.—Double carbonate of uranyl and thallium, ×300.

[1] Streng, Ber. d. oberh. Ges. xxii. p. 258.

bonate; only the crystals will then be smaller and not so beautifully formed.

§ 60. Chlorine

a. Precipitation with thallous nitrate.   Limit: 0.1 μgr. of Cl.

b. Precipitation with silver nitrate.   Limit: 0.05 μgr. of Cl.[1]

c. Precipitation as thallous chloroplatinate.   Limit: 0.004 μgr. of Cl.

d. Precipitation as potassium chloroplatinate.   Limit: 0.7 μgr. of Cl.

a. For the description of thallous chloride see § 6, *a*. If the test does not succeed, the delicacy may be at once raised more than a hundredfold by adding a trace of platinic sulphate (see § 60, *c*). If lead acetate be employed (§ 22, *a*) instead of thallous nitrate, the limit is reduced to 0.5 μgr. of Cl. The cubes of thallous chloride are, moreover, more characteristic than the needles of lead chloride.

b. As solvent Haushofer employs ammonia, while Streng uses hot hydrochloric acid. The latter has an advantage in some separations; but, on the whole, ammonia is preferable. See § 7, *a*.

c. Platinic sulphate is employed as the reagent. It must be added in small quantities to avoid crystallisation of double sulphates. Begin with test *a*, then, if necessary, add a trace of platinic sulphate. The crystals of thallous chloroplatinate are smaller than those of silver chloride, but still well developed.

d. Add a little platinic sulphate and a grain of potassium sulphate or a grain of nitre. Excess of platinic sulphate

[1] Haushofer, l.c. p. 117; Streng, Ber. d. oberh. Ges. xxiv, p. 54.

must be avoided, as in *c*. It might lead to the crystallisation of stout brownish prisms of platinic double sulphates. This test is less delicate than *a*, but the large yellow crystals of potassium chloroplatinate are easy to find. Its great value as a test lies in the possibility of distinguishing by its aid the three halogens Cl, Br, and I.

§ 61. Bromine

a. Precipitation with thallous nitrate.   Limit: 0.16 μgr. of Br.

b. Precipitation with silver nitrate.   Limit: 0.05 μgr. of Br.

c. Precipitation as thallous bromoplatinate.   Limit: 0.006 μgr. of Br.

d. Precipitation as potassium bromoplatinate.   Limit: 0.24 μgr. of Br.

e. Precipitation as thallous bromoaurate.   Limit: 0.7 μgr. of Br.

f. Staining of starch.   Limit: 2.0 μgr. of Br.

a. The crystals of thallous bromide resemble small crystals of the chloride (4 μ). They are less soluble, yet they may be recrystallised from a solution in hot water.

Lead bromide cannot be distinguished by the eye from lead chloride.

b. Silver bromide has the appearance of the chloride, but it is less soluble in ammonia; consequently, crystallisation from this solvent yields a less copious crop and crystals of only half the size.

c. Very small crystals, requiring a power of 400, otherwise not differing from crystals of thallous chloroplatinate.

d. Potassium bromoplatinate has a decided orange colour.

It crystallises before the chloroplatinate, chiefly around the centre of the drop, while the crystallisation of the chloroplatinate starts from the periphery. The same precautions must be observed as in § 60, *d*. Instantaneous reaction may be expected in solutions containing 0.3 per cent of potassium bromide.

e. If solutions containing 1 per cent of potassium bromide are tested with gold chloride and thallous nitrate, an orange-coloured powder of thallous bromoaurate is thrown down, dissolving at a gentle heat, and crystallising from the periphery of the drop in stout orange prisms. It is necessary to expel any excess of nitric acid, and to heat gently for recrystallisation, to avoid thallic chloride and reduction of gold. It is important also to add no more gold chloride than necessary, because excess would occasion crystallisation of thallous chloroaurate. Instantaneous reaction may be expected with 0.3 per cent of potassium bromide.

f. The yellow stain produced on starch by free bromine affords a convenient and characteristic test. The sample should contain a small quantity of a soluble chloride. Acidulate with sulphuric acid, add some grains of starch (no more than will cling to the point of a dry platinum wire), and a little potassium nitrite. If less than 0.2 per cent of potassium bromide be present, the sample must be covered with a small watch-glass for some minutes. The stain varies from pale yellow to orange with the quantity of bromine.

§ 62. Iodine

a. Precipitation with thallous nitrate. Limit: 0.17 μgr. of I.

b. Precipitation with silver nitrate. Limit: 0.17 μgr. of I.

c. Precipitation with palladium nitrate. Limit: 0.1 μgr. of I.

d. Precipitation as potassium iodoplatinate. Limit: 0.2 μgr. of I.

e. Precipitation as mercuric iodide. Limit: 0.2 μgr. of I.

f. Staining of starch. Limit: 0.17 μgr. of I.

a, b. These precipitates are marked by a light yellow colour. It is, however, difficult to distinguish iodine from bromine by their aid if the two halogens are mixed. Thallous bromide can be extracted with hot water, silver bromide with ammonia. Silver iodide is very sparingly dissolved; the solution will not yield distinct crystals. For distinguishing iodine from the other halogens the following tests are more suitable.

c. Dark brown precipitate, soluble in ammonia and in an excess of potassium iodide (§ 29, *a*). If the presence of bromine is suspected, chloride of palladium must be employed instead of the nitrate, as the latter reagent will produce in moderately diluted solutions of bromide a reddish-brown precipitate of $PdBr_2$.

d. With potassium sulphate and a little platinic sulphate dark gray octahedra of K_2PtI_6 are produced. They appear before the orange-coloured crystals of bromoplatinate. Instantaneous reaction is limited by a dilution of 1 part of potassium iodide in 800 parts of water.

e. An excess of mercuric chloride does no harm. The reaction is delicate as well as characteristic. As small crystals of mercuric iodide look black in transmitted light, a strong illumination from above should be employed. For more details see § 25, *a*.

f. The blue stain on starch is so intense that it generally passes in the course of half a minute from blue to black. In very dilute solutions a pale grayish-blue is produced. If too much concentrated sulphuric acid has been added this will modify the starch to a variety of dextrine, which is stained red by iodine. The test is managed as § 61, *f*, only no chloride is necessary. Hydrogen peroxide renders excellent service in liberating iodine; for bromine it is not suitable. If small quantities of bromides and iodides are to be traced in a great quantity of chlorides, evaporate, if necessary, with sodium carbonate (to decompose chlorides of Ca, Mg, etc.), and treat the dry mass with alcohol. The residue from the alcoholic solution is dissolved in water, and a drop is tested with platinic sulphate (§ 61, § 62, *d*). Another drop is tested with starch. A *small* quantity of nitrite facilitates the appearance of the blue stain of iodine; a greater quantity will not produce any indication of bromine if more than a trace of iodine be present, unless the stain of iodine be destroyed. But, as bromine is set free after iodine, it is often found in the liquid run off from the blue iodide of starch if this liquid be tested with some grains of starch. If necessary, small additions of hydrochloric acid and of potassium nitrite also are made. If bromine is not found return to the iodide of starch. It is covered with a thin layer of dilute hydrochloric acid, and a grain of potassium chlorate is added. The starch is bleached if no bromine is present; if bromine is present the colour of the starch is changed to yellow or orange, which resists the bleaching influence of chlorine for a long time.

§ 63. Fluorine

a. Precipitation as sodium fluosilicate. Limit: 2 μgr. of F.

b. Precipitation as barium fluosilicate. Limit: 0.7 μgr. of F.

a. The precautions enumerated in § 39, *a*, are here of no consequence. Sometimes short work can be made by dissolving silica in the acidulated sample on a varnished slide, and testing immediately with sodium chloride. At other times distillation must precede testing if the sample is charged with other salts. Silicates containing fluorine, such as topaz, must be prepared by fusion with sodium carbonate. The fused mass is nearly neutralised with sulphuric acid; it is acidulated with acetic acid, and evaporated in a platinum spoon. The dry residue is again moistened with concentrated sulphuric acid, and heated to 120° C., with another platinum spoon for cover. If distillation can be avoided, the limit is carried forward to 0.4 μgr. of F.

b. This test is not so convenient as the preceding one, but more delicate. Under favourable circumstances it will betray 0.15 μgr. of F. With rapid precipitation needles of 12 to 20 μ are produced. If it be possible the sample should be heated, when, on cooling, rods of 40 to 70 μ will separate out. For more detail, see § 19, *b*.

VI. TABLE OF REACTIONS

The numbers of the second column agree with those placed in the text beside the names of the elements. The numbers placed beside the formulæ give the limits of the reactions reduced to the unit of one micromilligram (1 μgr. = 0.001 milligr.) of the traced element.

Aluminium	42	Precipitation with $CsCl$: 0.35 ; with NH_4F : 0.3.
Antimony	49	Precipitation with $CsCl$: 0.08; with $C_2O_4H_2$: 1.0; with barium tartrate : 1.0 ; as antimonate of sodium : 0.5.
Arsenic	50, 51	Precipitation as As_2O_3 : 0.14 ; as ammonium-calcium arsenate : 0.035 ; as arsenomolybdate : 0.22.
Barium	19	Precipitation with H_2SO_4 : 0.05 ; as $BaSiF_6$: 0.09; as $BaCrO_4$: 0.08 ; as double tartrate of Ba and Sb : 0.45.
Beryllium	9	Precipitation with $K_2C_2O_4$: 0.08.
Bismuth	48	Precipitation with C_2O_4HK : 0.3 ; with $RbCl$: 0.13; with $HKSO_4$: 0.3 ; with KI and H_2O : 0.4.
Boron	41	Precipitation as KBF_4 : 0.2.
Bromine	61	Precipitation as $TlBr$: 0.16 ; as $AgBr$: 0.05 ; as Tl_2PtBr_6 : 0.006 ; as K_2PtBr_6 : 0.24 ; as $TlAuBr_4$: 0.7 ; as bromide of starch : 2.0.
Cadmium	14	Precipitation with $HNaCO_3$: 0.01 ; with $C_2O_4H_2$: 0.34 ; as double thiocyanate : 1.0 ; with $K_6Fe_2Cy_{12}$: 0.03.
Cæsium	4	Precipitation with $SnCl_4$: 1.6 ; as silicomolybdate : 0.25.
Calcium	21	Precipitation as sulphate : 0.04 ; as tartrate : 0.03; with K_4FeCy_6 : 0.015 ; with $C_2O_4H_2$: 0.06.
Carbon	40	Precipitation as $SrCO_3$ or as $PbCo_3$: 1.0.
Cerium	15	Precipitation with Na_2SO_4 : 0.02 ; with Na_2CO_3 : 0.05 ; with $C_2O_4H_2$: 0.04 ; with K_4FeCy_6 : 0.1.
Chlorine	60	Precipitation as $TlCl$: 0.1 ; as $AgCl$: 0.05 ; as Tl_2PtCl_6 : 0.004 ; as K_2PtCl_6 : 0.5.

Chromium	44	Precipitation as $AgCr_2O_7$: 0.025; as $PbCrO_4$: 0.02.
Cobalt	11	Precipitation as double nitrite: 0.1; as double thiocyanate: 0.3; as double phosphate: 0.02; as $Co_2Cl_6 10NH_3$: 0.2.
Copper	23	Precipitation as triple nitrite 0.03; with K_4FeCy_6: 0.1; as double thiocyanate: 0.1; as Cu_2I_2: 0.05.
Didymium	17	Precipitation with Na_2SO_4: 0.07; with Na_2CO_3: 0.1; with $C_2O_4H_2$: 0.1; with K_4FeCy_6: 0.15.
Fluorine	63	Precipitation as Na_2SiF_6: 2.0; as $BaSiF_6$: 0.7.
Gold	26	Precipitation with $SnCl_2$: 2.0; with thallous nitrate: 6.0.
Iodine	62	Precipitation as TlI: 0.17; as AgI: 0.17; as PdI_2: 0.1; as K_2PtI_6: 0.2; as HgI_2: 0.2; as iodide of starch: 0.17.
Iridium	30	Precipitation as Rb_2IrCl_6: 0.3.
Iron	43	Precipitation with K_4FeCy_6: 0.07; with NH_4F: 0.2; with barium oxalate: 0.1.
Lanthanum	16	Precipitation with Na_2SO_4: 0.04; with Na_2CO_3: 0.06; with $C_2O_4H_2$: 0.06; with K_4FeCy_6: 0.1.
Lead	22	Precipitation as $PbCl_2$: 0.3; as PbI_2: 0.2; as $PbSO_4$: 0.04; as $PbCO_3$: 0.06; as $PbCrO_4$: 0.10; as $Cs_2CuPb(NO_2)_6$: 0.03.
Lithium	3	Precipitation as $LiCO_3$: 0.36; as LiF: 0.25.
Magnesium	8	Precipitation as double phosphate ($NH_4MgPO_4 + 6H_2O$): 0.0012.
Manganese	10	Precipitation with $C_2O_4H_2$: 0.5; as double phosphate: 0.3; as MnO_2: 0.2.
Mercury	24, 25	Precipitation as Hg_2Cl_2: 0.25; as Hg_2CrO_4: 0.5; as HgI_2: 0.075; as double thiocyanate: 0.04.
Molybdenum	57	Precipitation as phosphomolybdate: 0.1; as Tl_2MO_4: 0.033.
Nickel	12	Precipitation as $K_2PbNi(NO_2)_6$: 0.008; as $NH_4NiPO_4 + 6H_2O$: 0.01.
Niobium	46	Precipitation as $NaNbO_3 + 3H_2O$: 0.6.
Nitrogen	57	Traced in nitrites with KI: 0.25 NO_3H; ammonia as $(NH_4)_2PtCl_6$: 0.1 NH_3; as $NH_4MgPO_4 + 6H_2O$: 0.05 NH_3; cyanogen as Prussian blue: 0.07 Cy.
Osmium	33	Precipitation with CsCl: 0.1; as $K_2OsO_4 + 2H_2O$: 0.1; as $N_4H_{12}OsCl_2$: 0.05.
Palladium	29	Precipitation with KI and NH_3: 0.1; with HCl and $TlNO_3$: 0.2; with NH_4CyS and $TlNO_3$: 0.07; as $N_2H_6PdCl_2$: 0.2.
Phosphorus	52	Precipitation as $NH_4MgPO_4 + 6H_2O$: 0.008; as ammonium phosphomolybdate: 0.015.
Platinum	27, 28	Precipitation with $CuCl_2$ and NH_3: 0.06; as K_2PtCl_6: 0.6; as Rb_2PtCl_6: 0.2; as Tl_2PtCl_6: 0.004.

Potassium .	1	Precipitation as K_2PtCl_6 : 0.5 ; as phosphomolybdate : 0.3 ; as $K_3Bi(SO_4)_3$: 0.2.
Rhodium .	31	Precipitation as double nitrite : 0.09 ; as oxalate : 0.4.
Rubidium .	5	Precipitation as silicomolybdate : 0.7 ; as chloroplatinate : 0.3.
Ruthenium	32	Precipitation with CsCl : 0.8 ; traced with NH_4CyS : 1.2.
Selenium .	55	Precipitation with Mg : 0.1 ; with KI : 1.0.
Silicon .	39	Precipitation as Na_2SiF_6 : 0.05 ; as silicomolybdate : 0.004.
Silver .	7	Precipitation as AgCl : 0.1 ; as $Ag_2Cr_2O_7$: 0.15.
Sodium .	2	Precipitation with acetate of uranyl : 0.8 ; as Na_2SiF_6 : 0.16 ; as $Na_3Bi(SO_4)_3$: 0.04.
Strontium .	20	Precipitation with H_2SO_4 : 0.2 ; as $SrCrO_4$: 0.8 ; as tartrate : 0.4 ; as $SrCO_3$: 0.4.
Sulphur .	54	Precipitation as $CaSO_4 + 2H_2O$: 0.2 ; as cæsium alum : 0.12 ; as $PbSO_4$: 0.006.
Tantalum .	47	Precipitation as K_2TaF_7 : 6.0 ; with NaOH : 1.2.
Tellurium .	56	Precipitation with Mg : 6.0 ; with CsCl : 0.3 ; with KI : 0.6.
Thallium .	6	Precipitation as TlCl : 0.16 ; as TlI : 0.03 ; as Tl_2PtCl_6 : 0.008.
Thorium .	38	Precipitation as sulphate : 30 ; with $C_2O_4H_2$: 0.1 ; with $(NH_4)_2CO_3$ and $TlNO_3$: 0.05.
Tin .	34, 35	Precipitation with $AuCl_3$: 0.07 ; with $HgCl_2$ 0.07 ; with $C_2O_4H_2$: 1.0 ; with CsCl : 0.45.
Titanium .	36	Precipitation as $K_2TiF_6 + H_2O$: 6.0 ; as $Rb_2TiF_6 + H_2O$: 1.0.
Tungsten .	58	Precipitation as WO_3 : 1.6 ; as phosphotungstate : 0.12 ; as Tl_2WO_4 : 0.08.
Uranium .	59	Precipitation with $C_2H_3O_2Na$: 0.6 ; with ammonium carbonate and $TlNO_3$: 0.1.
Vanadium .	45	Precipitation as NH_4VO_4 : 0.3 ; with $AgNO_3$: 0.07 ; with NO_3Tl : 0.07.
Yttrium .	18	Precipitation with $C_2O_4H_2$: 5 ; with oxalate and NH_3 : 0.03.
Zinc .	13	Precipitation with $NaHCO_3$: 0.01 ; with $C_2O_4H_2$: 0.1 ; as double thiocyanate : 0.1 ; with $K_6Fe_2Cy_{12}$: 0.05.
Zirconium .	37	Precipitation with C_2O_4KH : 0.06 ; as Rb_3ZrFl_7 : 0.5.

PART II

APPLICATION TO THE ANALYTICAL EXAMINATION OF MIXED COMPOUNDS

In writing these lines, I cannot but feel some apprehension that exaggerated expectations may have arisen in the minds of some readers who have followed me thus far. As I should be sorry to have damaged, by injudicious advocacy, a good cause, for which I have long laboured, I permit myself some remarks on this subject before proceeding further.

Many applications will be found for microchemical analysis which cannot be foreseen by me; this work therefore is probably a mere outline compared with manuals on the same subject that will be published twenty years hence, when the advantages of microchemical analysis will be understood everywhere, when its appliances will be fully developed, when difficulties have been surmounted, and obscurities have been cleared up. These are too numerous to be successfully dealt with by the small band of workers at present engaged in this new pursuit.

Having tried to work out a general method of examination, I have gradually come to the conviction that this would be a chimerical undertaking. My attempt in this direction has therefore been restricted to a summary instruction for separating the more important elements. Groups of the rare elements will be treated in separate chapters.

Some extensions that suggested themselves would have carried me beyond the limits of my experience. I should have been obliged to borrow facts and methods, not tried and verified by myself. Other applications have been put aside because they seemed to involve an unjustifiable trespass upon the domain of ordinary chemistry.

Some elements may be traced by microchemical methods to an extent which is quite beyond the reach of ordinary chemistry, but this extraordinary power of discovery is limited by several conditions. Take, for example, gold, dispersed in quartz rock or gravel, in the proportion of 1 to 1,000,000. Here we are on the domain of ordinary chemistry, working with great quantities. Perhaps even here an advantageous application of microchemical methods will be found, but this is a problem reserved for the future.

I. SYSTEMATIC SCHEME OF EXAMINATION

(1) Preliminary Tests

A. Given a liquid. Necessary tests—

1. The reaction has to be ascertained—if acid, neutral of alkaline.

2. The liquid has to be evaporated, and any residue noted.

3. Tests for volatile substances—HCl, NH_3, etc.—are to be applied.

§ 64. Testing for Reaction

1. Minute drops are put on test paper. If the sample is too small to admit any waste of substance, particles of a pasty dye-stuff, *litmus* or *congo red*, are made to touch a drop of the liquid. Litmus affords a more delicate test than congo red. This advantage is partly neutralised by the inconvenience of keeping and applying two test dyes —one blue, the other tinged a reddish-violet by adding a trace of nitric acid; whilst congo red, turned to a dirty violet by a trace of nitric acid, serves equally for acid and for alkaline liquids.[1] Litmus spreads rapidly, while congo dye-stuff is nearly insoluble in liquids of an acid reaction. In some cases the superior delicacy of litmus is more than counterbalanced by the impurities it always contains.

Carbonic acid, boric acid, traces of acetic acid are not indicated by congo dye-stuff. By some salts the behaviour of both colouring matters is altered.

Acetate of Barium.—Congo violet turns red. Litmus yields a blue precipitate, and small stars, turned red by acetic acid.

Acetate of Calcium.—Congo violet as above. With litmus a cloudy precipitate, turned red by acid.

Acetate of Magnesium. — Congo violet is somewhat

---

[1] An aqueous solution of litmus is evaporated over a water bath till it becomes pasty. Litmus extract of excellent quality is made by the "Societeit der Blauwsel-Fabriek Westzaan," at Westzaan, near Amsterdam.

Congo red is dissolved in boiling water. Three-quarters of the solution are mixed with dilute nitric acid until the solution is violet. If a decided bluish tint should prevail, boil with some of the reserved red solution. On cooling, violet flakes separate out. They are collected on a filter and washed with cold water till they begin to dissolve.

reddened.   Litmus behaves as with the calcium compound.

Acetate of Lead.—Congo violet : bluish-violet.   Litmus : a white cloud, turning speedily red.

Mercurous Nitrate.—Congo violet : violet, turning sluggishly blue.   Litmus : a thick whitish cloud, spotted with red.

Stannous Chloride. — Congo violet : violet, no blue. Litmus : red.

Ferric Chloride.—Congo violet : violet.   Litmus : immediately red.

§ 65. Testing for Volatile Substances

2. Testing for volatile substances may be performed on a slide in the following manner :—A drop of the sample is enclosed in a ring or triangle of glass or wire, which serves to support a short slide, touched with a drop of the reagent required for the test. As a rule, heat must be applied cautiously to avoid spirting.

To detect *carbonic acid* a mixture of ammonia and acetate of calcium or strontium (§ 40, *a*) may be used ; to detect *sulphuretted hydrogen*, acetate of cadmium.

Hydrochloric, *hydrobromic*, and *hydroiodic* acids are traced with salts of thallium or lead (§ 60, § 61, § 62, *a* ; § 22, *a*, *b*).   *Hydrocyanic* acid is not indicated by these reagents. With nitrate of silver it behaves like hydrochloric acid. Nothing remains but to fix it with caustic alkali and to apply the test for Prussian blue (§ 53, *c*).

Fig. 70.—Distilling tube for microchemical tests.   Full size.

For distilling hydrocyanic acid small bulb-tubes are convenient.[1] Through A the liquid to be distilled is introduced, a small drop of water is made to follow it, and the point A sealed with the blowpipe flame. The preparation is completed by inserting through B a thread of glass or a platinum wire dipped in potash lye.

Nitrous Acid is traced by means of potassium iodide and starch (§ 53, *a*). *Chlorine* will give the same reaction; by excess of chlorine it will be destroyed. *Nitric acid* requires the addition of a trace of sodium chloride. As a specific reagent for nitric acid, acetate of *cinchonamine* ($C_{19}H_{24}N_2O \cdot HNO_3$) may be used.[2] The nitrate of this base has about the same solubility as calcium sulphate; it crystallises readily, yielding colourless, rectangular, and six-sided plates (rhombic system), measuring from 25 to 120 μ.

FIG. 71.—Nitrate of cinchonamine, ×130.

Sulphurous Acid may be traced by its reducing action on a mixture of potassium ferricyanide and acetate of uranyl, acidulated with acetic acid. The liquor, drained off from the red precipitate, is oxidised with nitric acid or with potassium chlorate and hydrochloric acid, and tested for sulphuric acid (§ 54, *a*, *b*).

Ammonia, ammonium carbonate and sulphydrate may be distilled without difficulty. Ammonia is fixed with hydrochloric acid and tested with platinic chloride. Fix-

---

[1] Such tubes are made by drawing out a narrow tube at *c* and *d*,

FIG. 72.—Distilling tube, half finished. Full size.

blowing a bulb from the thicker piece *cd* and repeating these manipulations at *ef*. Finally *de* is drawn out and bent in a small Bunsen flame.

[2] Arnaud, Ann. chim. et phys., 1890, tom. xix. p. 123.

ing and testing can be combined by the use of phosphomolybdic acid (§ 53, *b*).

Testing for sulphur in ammonium sulphydrate is rather difficult, the amorphous precipitate of cadmium sulphide being easily masked by hydroxide and carbonate. A good reagent for this purpose is yet wanting.

§ 66. Evaporation Test

3. On being evaporated at a gentle heat drops of dilute solutions generally leave a sharply drawn, circular, or elliptic line, marking the original boundary of the drop. When not perceived by the naked eye, this line is often distinctly seen under a pocket-lens or a low power of the microscope. The delicacy of the test is increased by making up the volume with some of the original solution, when the evaporating drop is diminished to half its height, and repeating this manipulation several times.

Causes of Error

a. Distilled water is rendered impure by prolonged contact with glass, alkali and silica being dissolved. Water that has been put aside for some hours in a pipette, drawn from a narrow tube of soft glass, has taken up so much impurity that a single drop will leave a distinct stain. Hard glass resists a longer time, but after some days the same amount of impurity will be found.

For ordinary work a correction is possible. Evaporate a drop of water close to the drop that has to be tested and compare the residues.

For exact work the water must be distilled on the spot. Bend small test-tubes to an angle of 120°, then hold the open end in the flame till it shrinks. Distilled water is poured in and heated over a small Bunsen flame, the upper half of the tube being held in a nearly horizontal position.

FIG. 73.—Tube for distilling water, acids, etc. Half size.

The first drops are used for rinsing the glass rod or pipette; the fourth drop may be taken as sufficiently pure. These remarks apply equally to the testing of reaction (§ 64).

b. By hydrochloric acid, caustic ammonia, and ammonium chloride even hard glass is strongly attacked. The corrosive action of nitric and sulphuric acids is less marked, yet by no means to be neglected. For delicate tests, impurities taken up in the bottles may be eliminated by distilling a few drops of the acids, and a coating of hardened Canada balsam protects the glass of the slides against the corroding action of acids, but not unfrequently these precautions fail. Indeed, the heating and evaporating of strong acids on glass slides is always to be looked upon as involving a source of error.

No material is known that will meet all requirements. Thin plates of quartz are clear and exempt from colour; they resist corrosion by common acids perfectly well, but are spoiled by caustic alkali and by soluble fluorides. Platinum foil (thickness about 0.2 mm.) can be used only with reflected light, and must be preserved from the action of free halogens. It must be kept bright and free from creases and irregular scratches. The fine parallel lines left on the foil by the finishing rolls are of no consequence.

For cleaning, a soft cork is convenient, the surface of which is from time to time renewed by paring with a sharp knife. If burnishing is necessary, this is done with rouge or strongly calcined oxide of chromium.

B. Given a solid substance. On this is to be tried—
1. Solution.
2. Sublimation.

§ 67. Solution

1. With *metals* try at once nitric acid; if the desired effect is not obtained try aqua regia. For details see the section on Examination of Alloys.

If the substance has not the appearance of a metal it is heated with water. If anything is dissolved (see Evaporation Test, § 66) the treatment is repeated three or four times. The last portion of the aqueous solution is generally thrown away.

On a particle of the residue the effect of nitric acid should be tried, on another particle that of hydrochloric acid, beginning with dilute acids, and taking the solvent that has given the best effect for the treatment of the whole residue. Care should be taken to observe if bubbles of gas arise, or a particular odour is given off (see § 65). Any residue is to be heated with water, because some nitrates and many chlorides are but sparingly soluble in strong acid. Boiling with acids should be performed on platinum (§ 66, *b*) or porcelain.

A particle of the residue left by hydrochloric or nitric acid is tried with aqua regia, another with strong sulphuric acid (afterwards with water). By heating with sulphuric acid, fluorides, fluosilicates, fluotitanates, etc., are decomposed; from insoluble cyanides hydrocyanic acid is expelled (under certain circumstances carbon monoxide). Insoluble oxalates are converted to sulphates, the oxalic acid being entirely broken up to monoxide and dioxide of carbon.

Some refractory compounds, as aluminium oxide, titanic and zirconic acids, spinel, tungsten, columbite, are decomposed by fusion with hydrogen potassium sulphate.

Other insoluble compounds—sulphates, phosphates, the majority of silicates, chloride of silver, oxides of antimony and tin—are prepared for dissolving in acids by fusion with four times their weight of sodium carbonate and subsequent treatment with water.

§ 68. Sublimation Tests

Iron or nickel wire of 0.3 to 0.5 mm. diameter is cut into lengths of 10 cm. The ends are flattened under the hammer, cleaned by scraping or filing, and oxidised over a Bunsen flame. One or two milligr. of substance made into a paste with water or hydrochloric acid are put on the wire, and it is heated in the point of a bright flame of about 15 mm. length, at a distance of 5 mm. from the sample. A slide, held in the left hand, is made to rest against the wire, the end of the latter being kept about 2 mm. under the glass. The wire is then drawn on with a steady motion, so that its end passes over the point of the flame. To ensure success the sample must receive sufficient heat, while the slide is comparatively cool. If the glass becomes too warm or is held at too great a distance over the sample, the films will spread over too great a breadth and be nearly useless.

§ 69. Sublimation of Oxidation Products

The substance is prepared by heating with nitric acid

until cessation of nitrous fumes, evaporating and moistening with water. From samples of this kind the following elements and compounds may be obtained in films:— *sulphur, selenium, selenium dioxide, arsenious acid, mercury, tellurium, tellurium dioxide, antimonious oxide.* They are arranged in the order in which they are driven off.

Sulphur is mostly burnt. A small quantity is condensed in small yellow drops.

Selenium Dioxide may be reduced by sulphur. Rapidly condensed selenium gives a red powdery film, the red colour of which is especially vivid on the under-side. Where the glass has got warm, selenium collects in small black drops. Part of it is burnt to dioxide, forming a whitish film which is stained red by an acidulated solution of $SnCl_2$.

Arsenious Acid yields readily white, fine-grained films. A small drop of hydrochloric acid will penetrate to the glass, *lifting the film* which will float awhile, becoming puckered and somewhat crystalline before breaking up and sinking.

Mercury is always condensed in the metallic state, even when sulphide of mercury is subjected to sublimation. It is widely dispersed, looks coarse-grained and black when illuminated from below, while in reflected light the dark field is speckled with white shining beads. A delicate test, highly characteristic.

Tellurium will sometimes yield metallic films, almost opaque, of a brownish-gray. It is for the most part burnt to white dioxide, transmitting a brownish tint. It is easily dissolved in hydrochloric acid; cæsium chloride precipitates yellow octahedra from this solution (§ 56, *b*).

Antimonious Oxide requires a dark red heat for sublimation. The films are very fine-grained, almost continuous, reflecting a bluish tint. They dissolve *instantly* in hydrochloric acid, *without floating and wrinkling*. For further discrimination between As and Sb the solution should be touched with a platinum wire dipped in a solution of potassium iodide. If only As is present, the solution is not tinged; a bright yellow precipitate (AsI_3) is formed, dissolving in the acid and renewed where it touches the film. From its solution in dilute hydrochloric acid, the tri-iodide of arsenic crystallises in thin, six-sided plates, and groups of slender rods of a bright canary yellow.

If only Sb is present, the solution takes a yellow tint; where it touches the film orange-coloured dendrites are slowly formed. If both As and Sb are present, a yellow liquid and a yellow precipitate are produced, the orange-coloured dendrites of SbI_3 appearing much later, when the drop of acid is nearly dried up.

FIG. 74.—Arsenious iodide, × 200.

Arsenious Acid and *Selenium Dioxide.*—Add a small drop of hydrochloric acid, a minute later touch with potassium iodide. If Se is present the liquid turns yellow and red SeI_4 separates out, masking the arsenic tri-iodide. Let the drop become nearly dry, heat with water, and concentrate till a yellow seam of AsI_3 appears. The same treatment is applied to the combination $As_2O_3 + TeO_2$ (see § 56, *c*). Begin by testing for TeO_2 with cæsium chloride (§ 56, *b*).

§ 70. Sublimation of Chlorides

The mixture of oxides and nitrates (§ 69) is calcined to incipient redness, then evaporated to a paste with hydrochloric acid. For rough work, the residue from the sublimation of oxides is moistened on the wire with hydrochloric acid, dried, moistened once more with acid, and subjected to sublimation.

In this way films can be obtained of $CdCl_2$, $BiCl_3$, $ZnCl_2$, $CuCl_2$, $PbCl_2$.

Cadmium Chloride is very easily driven off. The film is white except where reduction to metal and oxide has taken place. Here its colour is a yellowish-brown. It may be tested with potassium oxalate (§ 14, *b*).

Bismuth Chloride sublimes easily without decomposition. Its solution in dilute hydrochloric acid is coloured intensely yellow by potassium iodide. It may be tested with acetic acid and potassium oxalate, or with potassium sulphate (§ 48, *a*, *c*).

Chloride of Zinc.—The chloride is deposited first, and later the oxide, dissolving slowly in dilute acids. It should be evaporated with a small drop of hydrochloric acid, and tested with sodic carbonate (§ 13, *a*).

Cupric Chloride sublimes at first easily, giving a brownish film. When decomposition has set in (at a higher temperature) more heating is required for the sublimation of white cuprous chloride. Testing is effected with potassium nitrite and acetate of lead (§ 23, *a*) or with thiocyanate (§ 23, *c*). In the latter case addition of zinc acetate is very advantageous.

Chloride of Lead requires a strong heat for sublimation. The films are grayish-white, granular, changing to an aggregate of fine needles when moistened by the breath. After heating with a drop of water, the common forms of $PbCl_2$ are obtained (§ 22, *a*). A trace of potassium iodide produces glittering flakes of PbI_2 (§ 22, *b*).

Stannic Chloride is not condensed in films.

Ferric Chloride is sparingly sublimed when strongly predominant in the sample.

Ferrous Chloride and the chlorides of *cobalt* and *nickel* do not yield any films.

To isolate Zn and Cd by sublimation, the greater part of the heavy metals must be thrown down upon bright iron from a nearly neutral solution in hydrochloric acid, the liquid evaporated, and sublimation proceeded with.

Even when calcining has been duly carried out, some arsenic and antimony may find their way into the films of chlorides, masking lead and bismuth. *Lead* may always be traced by treatment with water and potassium iodide (no acid is to be used, and only a *trace* of potassium iodide). Bismuth is traced with potassium sulphate. If antimony is absent, bismuth, sublimed along with lead, is betrayed by the yellow colour produced by hydrochloric acid in the drop from which lead iodide has separated out.

§ 71. Sublimation of Water

Testing for water is quite easy, when at least half a milligram is to be counted on. A narrow tube is drawn out as shown below.

The sample is placed at *b* and dried by heating to 110° C. and sucking at *e*. After this the tube is cut at *d*, drawn

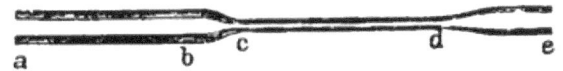

FIG. 75.—Testing-tube for sublimation of water. Full size.

out and sealed between *a* and *b*. It is then seized near *c* in a pair of forceps, and heated at the thick end in a small Bunsen flame. By judicious management the water can be accumulated in a narrow ring near *c*, or collected as a short column in the capillary part of the tube for microchemical examination (§ 65).

For minute quantities, the testing-tubes must be half the size of the drawing. Bore of the thicker parts 1 to 1.5 mm., of the thinner part 0.2 to 0.3 mm.; length from a to c 25 to 30 mm.

Drying in such small tubes is done by holding the thicker part in heated forceps, while air is drawn through.

If less than 0.1 mgr. of water is to be traced an indicator becomes necessary. Coal-tar dyes that are easily dissolved in water, such as methylene blue or malachite green, may be utilised for this purpose. A thread of glass or a thin wire, dipped in the fine powder, is inserted in the narrow part and left there during the operation. By this device we are enabled to detect 0.03 mgr. of water, but it labours under two defects—viz. the indicator is not dried in the testing-tube, and the results of the operation cannot be judged before the wire is withdrawn.

A thin coating of potassic permanganate in the capillary part of the tube answers these requirements, excepting in some cases, when reducing substances are volatilised with the water. The coating is applied by placing a grain of permanganate and a minute drop of water at d, heating gently and sucking at a.

The dry coating looks gray, wetted it looks intensely violet. If the dusty gray persists near the open point, no water can possibly have entered from without during the experiment. The delicacy equals that of coal-tar dyes (limit: 0.03 mgr. of water).

(2) EXAMINATION IN THE WET WAY

§ 72. Given an Aqueous Solution

a. Note the reaction. If it be alkaline, it must be

remembered that volatile compounds (CO_2, H_2S, HCy, etc.) may be driven out by adding acids, and that in an alkaline solution many compounds may be present, which are insoluble in dilute acids ($PbSO_4$, $PbCrO_4$, SiO_2, WO_3, As_2O_3, Sb_2O_3, Nb_2O_5, Ta_2O_5, TeO_2, several cyanides, sulphides, etc.)

b. If it be neutral, the majority of metals cannot be present. Test a small drop for halogens (§ 60, *b*), excluding silver, and for sulphuric acid (§ 54, *a*) by which lead, barium, and strontium are excluded.[1]

c. If the solution shows acid reaction, test a drop for halogens, for sulphuric, phosphoric, and arsenic acids (§ 51, *b*).

§ 73. Given a Solution in Nitric Acid—Precipitation of Oxides and Basic Nitrates

Test a portion for ammonia (§ 53, *b*). This portion is put aside to be tested for acids (§ 78).

Test a drop for mercurous salts (§ 24, *a*) and for phosphoric and arsenic acids (§ 51, *b*).

a. Heat the sample to ebullition with some drops of nitric acid, keep the heat up for about two minutes, precipitating TIN as dioxide (§ 35, *a*).

b. Evaporate till nearly dry, make up the original bulk with water. If a precipitate is formed, add an equal volume of water and apply heat to facilitate settling. The precipitate may contain basic nitrates of *bismuth* and *mercury*, also *telluric dioxide*. If phosphoric and arsenic acids are present, they accumulate in this precipitate and

---

[1] If K or Na are present, also Bi and the metals of the cerium group are excluded.

in the stannic dioxide (see § 79). Test for Bi and Te with hydrochloric acid and cæsium chloride (§ 48, *b*; § 56, *b*); for mercury with ammonium thiocyanate and cobaltous nitrate (§ 25, *c*).

§ 74. Precipitation of Chlorides and Iodides

a. To the solution resulting from § 73, *b*, hydrochloric acid is added. *Silver* is precipitated, *thallium* from solutions that are not highly diluted,[1] *lead* from moderately concentrated solutions.

The precipitate is heated with water. *Thallous* chloride and *lead* chloride are dissolved and distinguished by crystallisation (§ 6, *a*; § 22, *a*).

Any residue is treated with caustic ammonia, which dissolves *silver* chloride (§ 7, *a*). A gray or blackish tint indicates that some mercurous nitrate had escaped oxidation (§ 24, *a*).

b. In the liquid drained off from the precipitated chlorides small quantities of Bi and Sn may be present. A drop is concentrated and tested with cæsium chloride (§ 35, *a*; § 48, *b*).

After this, evaporate the solution, dissolve in water, add a little acetate of ammonium and ammonium iodide. A red precipitate indicates *mercury* (§ 25, *a*).[2] If *copper* be present in any considerable proportion, the liquid becomes yellow, iodine being set free. Drive it off by heating, then add a drop of hydrochloric acid. A yellow colour in the liquid points to *selenium*. Selenium tetraiodide is precipitated slowly in reddish flakes (§ 55, *b*), beginning with

---

[1] Traces of Tl are detected with $PtCl_4$ (§ 6, *c*).

[2] Sometimes accompanied or closely followed by glittering yellow scales of PbI_2.

a smoky seam round the drop. As cuprous iodide may be precipitated along with it, the precipitate is dissolved in hydrochloric acid with the aid of potassium chlorate, free chlorine driven off, and a part of the solution tested for *copper* (§ 23, *a*, or *c*), the rest for *selenium* with stannous chloride.

§ 75. Precipitation of Carbonates

To the solution which has been treated with ammonium iodide (§ 74, *b*) add ammonia and ammonium carbonate.

a. Precipitated: manganese, iron, chromium, aluminium, barium, strontium, calcium, and a small quantity of lead. Chromic and molybdic acids form insoluble compounds with Ba and Pb, also phosphoric and arsenic acids may occur, united with aluminium and iron.

The precipitate is dissolved in a drop of nitric acid, *chromium* oxidised to chromic acid, and *manganese* to peroxide (§ 10, *c*) by boiling with potassium chlorate. The acid liquid is tested for *chromic* acid with acetate of lead (§ 44, *b*).

b. Excess of lead is removed as iodide, then one portion is tested for *calcium* with sulphuric acid (§ 21, *a*), for *aluminium* with cæsium chloride (§ 42, *a*), for *iron* with potassium ferrocyanide (§ 43, *a*).

c. Another portion is tested for *barium* with ammonium fluosilicate (§ 19, *b*), for *strontium* with potassium bichromate.

For another method of distinguishing Ba, Sr, Pb, Ca, see p. 165, § 83.

§ 76. Precipitation of Oxalates

Add oxalic acid to the ammoniacal solution (§ 75, *a*) and acidulate with acetic acid.

a. Zinc, cadmium, cobalt, nickel, and traces of *copper* are precipitated.[1]

The oxalates are heated with sulphuric acid until white fumes are given off. Excess of caustic soda extracts the oxide of *zinc*, which is detected by adding carbonate of ammonia to the alkaline solution (§ 13, *a*, and *b*).

b. The residue is dissolved in acetic acid. *Cobalt* is precipitated with potassium nitrite (§ 11, *a*), *nickel* with potassium nitrite and acetate of lead (§ 12, *a*); or, cobalt and nickel are precipitated with ammonia and sodium phosphate (§ 11, *c*; § 12, *b*), the precipitate dissolved in acetic acid and tested with nitrite. An error, that might arise from traces of copper, is in this way avoided.

c. The liquid resulting from *b* may contain cadmium and copper.

Concentrate, acidulate with acetic acid, test for *cadmium* with oxalic acid (§ 14, *b*), for *copper* with potassium nitrite and acetate of lead (§ 23, *a*), or at once for both metals with thiocyanate of ammonium and mercury (§ 14, *c*).

§ 77. Separation of Alkali Metals from Magnesium

The solution resulting from § 76, *a*, is evaporated, the residue ignited and treated with hot water.

---

[1] *Uranium* essentially modifies the precipitation of several oxalates. Its presence is established by testing a small drop of the ammoniacal solution with thallous nitrate (§ 59, *b*). Evaporate and treat with caustic ammonia, which does not dissolve the uranium compound.

a. The residue may contain *beryllium* and *magnesium*. Dissolve in hydrochloric acid, drive off the excess, add caustic alkali and a drop of water, evaporate and treat with water. The alkaline solution, containing beryllium, is evaporated with ammonium chloride, the residue washed with water, dissolved in hydrochloric acid, and tested for *beryllium* (§ 9) after expulsion of an excess of acid. For small quantities addition of a little acetate of zinc is advantageous.

Magnesium is found (§ 8, *a*) in the residue from the treatment with caustic alkali.

b. The aqueous solution is evaporated. Extraction with cold water leaves carbonate of *lithium* behind. It is dissolved in a small drop of hydrochloric acid. Expel the excess of acid, dissolve in a small drop of water and test with ammonium fluoride (§ 3, *b*).

c. Evaporate once more, to judge if the residue allows splitting in two portions. If this is the case, test for *sodium* in one portion with acetate of uranyl (§ 2, *a*, *b*), and in the other portion for *potassium* with platinic chloride (§ 1, *a*). If splitting is not feasible, test with ammonium fluosilicate or with sulphate of bismuth (§ 2, *c*, *d*).

§ 78. Examination for Acids

a. This examination is shortened by the testing for phosphoric and arsenic acid, already done (§ 73, *a*), and by conclusions, drawn from the tests for metals, hydrochloric acid being excluded by the presence of silver and mercurous compounds; sulphuric acid by the presence of barium, strontium, lead; chromic acid by lead, etc.

b. In the portion that has been tested for ammonia (§ 73, *a*), chromic and vanadic acid may be traced by precipitating *vanadic* acid with excess of ammonium chloride (§ 45, *a*), *chromic* acid with acetate of lead (§ 44, *b*). The slender needles of the chromate of lead may be recognised amongst the plumbic compounds of several other acids.

c. Molybdic and *tungstic* acid are best precipitated from strongly acidified solutions with an excess of ammonium salts and a trace of sodium phosphate (§ 57, *a* ; § 58, *b*). If tungstic acid is present along with molybdic acid, nothing remains but trying to isolate it by means of nitric acid (§ 58, *a*).

d. Boric and *silicic* acid may always be separated from metals and other acids by distilling with ammonium fluoride and sulphuric acid (§ 39 ; § 41, *a*); *boric* acid also by subliming with ammonium fluosilicate (§ 41, *b*).

§ 79. Elimination of Phosphoric and Arsenic Acid

In many cases the presence of these acids is objectionable, because the behaviour of many metals is altered by them, and because they appear in almost every stage of the examination.

a. To get rid of them, *ammonium tungstate* may be used or *stannic oxide*—the *tungstate* only for small quantities. It is dissolved in dilute ammonia before adding to the sample, which must be strongly acidified with nitric acid. On boiling, insoluble phosphotungstate and arsenitungstate of ammonium are produced. Care must be taken to have a sufficient quantity of ammonium salts in the liquid, lest soluble phosphotungstic and arsenitungstic acids be produced, and to boil and evaporate with excess of nitric acid. If the operation has been conducted with due care, dilute nitric acid dissolves the metals free from phosphoric, arsenic, and tungstic acids. In dubious cases the operation must be repeated with a smaller dose of ammonia and ammonium tungstate. Turbid solutions are cleared by evaporating and treating the residue with dilute nitric acid.

b. For greater quantities, treatment with *tin* and nitric acid is to be preferred. Add small shreds of pure tin-foil (four times that of the probable quantity of phosphoric acid) to the strongly acidulated sample. After some minutes apply heat and add strong nitric acid, to convert all the tin to insoluble dioxide. Evaporate and extract the metals with hot water (bismuth with dilute nitric acid). When dealing with *phosphoric* acid, this method ensures complete success. When *arsenic* acid is present, this compound is partly reduced to *arsenious* acid, somewhat soluble in water and soluble to a considerable extent in dilute acids. As it is apt to mask several tests (*e.g.* the precipitation of plumbic iodide), evaporate and heat on platinum-foil until the arsenious acid is sublimed.

§ 80. Separation of the Alkali Metals (K, Na, Li, Rb, Cs, Tl)

a. If necessary, the solution is freed from sulphuric acid by means of barium chloride, an excess of which is removed with ammonium carbonate. It is then evaporated with hydrochloric acid and the residue heated (under cover) to expel ammonium chloride. While yet slightly warm, the residue is washed with alcohol. For very small samples amylic alcohol is to be recommended as being less volatile and not apt to absorb water from the surrounding air.

Collect the alcoholic solution in a corner of the slide and heat the glass at some distance above it, to prevent creeping and spreading out. If amylic alcohol has been employed, the residue must be heated to 140° C., in order to drive off heavy alcohols, which might be troublesome by making the aqueous solution milky.

b. Dissolve in a drop of water, test with a trace of sodium iodide for *thallium* (§ 6, *b*), then put one-half of the liquid on a varnished slide, where it is tested for *lithium* with ammonium fluoride (§ 3, *b*).

To detect rubidium and cæsium evaporate the rest, bring the residue in contact with a drop of a saturated solution of potassium chloroplatinate (K_2PtCl_6), and cover with a small watch-glass. Chloroplatinate of *cæsium* is immediately thrown down in yellow crystalline grains, measuring from 3 to 5 μ (the thallous compound, nearly insoluble, 1 to 2 μ). In about two minutes well-defined crystals of *rubidic* chloroplatinate turn up, measuring from 10 to 15 μ. They spread over a considerable distance in the drop of potassium chloroplatinate.

c. A portion of the residue from the extraction with alcohol is put into a drop of a solution of platinic chloride (1 : 50). Yellow octahedra that may reach 70 μ indicate potassium (§ 1, *a*). The potassic chloroplatinate is kept dissolved for a long time, when much sodium is present; it is therefore prudent to leave the drop to spontaneous evaporation. When the residue is moistened by breathing over it, even traces of potassium become distinctly visible.

To detect a minute proportion of *sodium* is attended with greater difficulty. The reaction (§ 2, *a*), otherwise to be recommended for its elegance and rapidity, fails under these conditions. Addition of magnesium acetate or zinc acetate may be turned to account (§ 2, *b*). A saturated solution of potassium antimonate is very serviceable. Add some grains of the dry mixture of potassium and sodium chlorides and hasten the crystallisation by adding a small drop of alcohol (§ 49, *d*).

§ 81. Solutions containing Hydrochloric Acid

Silver is excluded. *Mercury* may be present, and likewise *thallium* if aqua regia has been employed as solvent. Both are precipitated by stannous chloride.

From hydrochloric solutions that do not contain an excess of acid, oxychlorides of *bismuth*, *antimony* and *tellurium* are precipitated by water.

Stannic oxide is precipitated by boiling, especially when water and ammonium nitrate have been added. *Tin* and *tellurium* are easily recognised by means of cæsium chloride (§ 35, *a*; § 56, *b*).

Titanium dioxide is precipitated with the dioxide of tin. It is stained a brownish-yellow by potassium ferrocyanide and dilute hydrochloric acid.

To establish the presence of tin reduce on zinc, adding a drop of hydrochloric acid, wash the metallic powder on to a slide, dissolve in hydrochloric acid, and test with oxalic acid (§ 34, *c*), or with chloride of gold (§ 34, *a*).

Zirconium dioxide is not precipitated. See Rare Elements, § 152, § 153.

a. Much difficulty is to be expected when a small quantity of *antimony* has been precipitated with a great quantity of bismuth.

The safest way is to test for *bismuth* with potassium oxalate or potassium sulphate (§ 48, *a*, *c*); after this to fuse with the fivefold volume of potassium nitrate till the bead is reduced to half its size. Lixiviate with hot water, concentrate and precipitate with sodium chloride (§ 49, *d*).

b. Another method requiring some caution is based on the solubility of antimonious oxide in *tartaric acid*. The oxychlorides are heated with water and a little tartaric acid. Antimony and a small quantity of bismuth are dissolved, the latter as chloride. Evaporate, heat with a big drop of water, and drain off. A film of bismuth oxychloride will be left behind. Concentrate, add a small drop of hydrochloric acid, test with cæsium chloride, and a trace of potassium iodide (§ 49, *a*). A yellow colour of the liquid and orange-red crystals indicate *antimony*.

Testing with barium compounds (§ 49, *c*), as proposed by Streng, is not to be recommended. The proportions must be balanced to a nicety if the test is to succeed.

§ 82. Analytical Examination of Sulphates

Solutions containing sulphuric acid result from samples of rocks, decomposed with hydrofluoric and sulphuric acids, or from fusion of some of the more rare minerals with hydrogen potassium sulphate. Both subjects are to be treated in separate sections (§ 108, § 151).

In this place only one prominent feature of solutions in sulphuric acid claims immediate attention, viz. the frequent occurrence of residues, sometimes of a perplexing nature. Many sulphates are apt to unite in binary compounds, sparingly soluble in water and in dilute acids, such as the double sulphates of bismuth, zirconium, thorium, and of the cerite metals, with those of the alkali metals, and with those of several bivalent metals. Other sulphates, especially those of the barium group, occurring frequently in residues will take up various impurities, sulphates, nitrates, and chlorides, thereby becoming greatly modified as regards their form and colour.

Thus, for instance, the sulphates of barium and strontium take a yellow tint if precipitated from solutions containing much ferric chloride. At the same time, their solubility is enhanced and the relation of solubility modified. All varieties of crystals of barytes and celestine may be imitated in this way, the barium sulphate appearing in crystals quite as large as those of the strontium sulphate (20 to 40 μ). Under the same circumstances, calcium sulphate is modified to such an extent that it is sometimes entirely overlooked. Its characteristic oblique rods and twins (§ 21, a) are reduced to insignificant squares and grains (10 to 20 μ), that may be taken for sulphate of strontium or barium. Similar effects are produced by the chlorides of aluminium and chromium. Boiling with a dilute solution of ammonium acetate precipitates sesquioxides and restores the ordinary forms of calcium sulphate.

§ 83. A Method for distinguishing the Sulphates of the Barium Group (Ba, Sr, Pb, Ca).

Wash the mixed sulphates rapidly, then heat with a large drop of water. Drain the liquid off and concentrate. Crystals of gypsum will separate out, growing to a considerable size if the slide is put aside for half an hour. Meanwhile the treatment with water is repeated two or three times, with a view to concentrate the less soluble sulphates, at the same time thoroughly washing out alkaline sulphates and free sulphuric acid. Then boil with pure hydrochloric acid of sp. gr. 1.12, drain off and concentrate. Sulphate of strontium crystallises first; it is followed by sulphate of lead. In the end long needles of gypsum separate out. A second boiling with hydrochloric acid usually yields sulphates of strontium and lead, accompanied by the far greater rhombs and rods of the chloride of lead (§ 22, *a*).

Strontium sulphate appears under these circumstances in squares and rectangles of 4 to 10 μ, *lead sulphate* in oblong hexagons, measuring from 12 to 30 μ. The crystals of the *chloride* of lead are thrice as large as those of the sulphate. When they appear, add a trace of potassium iodide (§ 22, *b*). If the sulphate of strontium should be masked by a great excess of the plumbic compounds, the latter may be extracted with caustic alkali and the lead precipitated as carbonate (§ 22, *d*) by means of sodium bicarbonate.

FIG. 76.—Strontium sulphate, ×200.

Of the *barium* sulphate only traces are dissolved in hydrochloric acid of moderate strength. When lead and the greater part of the strontium have been removed, mix the dried residue with thrice its volume of sodium carbonate, fuse on a platinum wire, lixiviate with water, dissolve in hydrochloric acid, and test for *barium* with ammonium fluosilicate (§ 19, *b*).

§ 84. Treatment of the Mixed Sulphates of Bismuth, Calcium, and Sodium

If sulphate of bismuth be present, the rods and needles of gypsum are associated with grains and disks that may be confounded with the double sulphate of bismuth and potassium (§ 1, *c*). New complications arise through the presence of sodium. Even a practised eye will not be able to distinguish small crystals of gypsum from those of the double sulphate of bismuth and sodium (§ 2, *d*). The difficulty cannot be solved by heating with water; and recrystallisation from a solution in hydrochloric acid is equally useless. The bismuth is most readily thrown down as reddish oxyiodide (§ 48, *d*) by heating with hydrochloric acid and adding ammonium iodide and water. From the yellowish solution normal crystals of gypsum may be obtained by concentration. Sodium is easily traced after evaporation and calcination of the residue (§ 2, *a*). Complications, arising from the presence of the cerite metals, will be treated in the section on the rare elements (§ 155).

II. APPLICATION OF MICROCHEMICAL ANALYSIS TO EXAMINATION OF WATER

§ 85. Qualitative Water Analysis

For a preliminary examination of water, microchemical analysis has a great advantage in saving time and dispensing with cumbrous and complicated apparatus. About 40 cub. centim. of water will generally suffice for ordinary examination, and such an examination, evaporation of the water included, may be performed in about two hours.

Suppose that the water is to be tested for potassium, sodium, calcium, magnesium, iron, lead; for chlorides, phosphates, sulphates, and carbonates; for ammonia and nitrous acid.

Then one portion (*a*) of about 20 cub. centim. is to be concentrated by evaporation to 1 cub. centim.; another portion (*b*) of 10 cub. centim. is subjected to the same treatment after addition of two drops of nitric acid; a third portion (*c*) is treated in the same manner after addition of a little caustic alkali.

a. A drop of the concentrated liquid (*a*) is acidulated with nitric acid and subjected to further concentration on a slide. *Sulphate of calcium*, if present, will be immediately detected (§ 21, *a*).

To the same drop thallous nitrate is added to detect *chlorides*.

Two or three small drops are evaporated on the same spot, with addition of a grain of calcium acetate. The residue is covered with a layer of a tepid solution of gelatine and the slide put aside in a cool place for a quarter of an hour. After that time a drop of hydrochloric acid is spread over the jelly, which will speedily penetrate it, evolving bubbles of *carbonic acid* if soluble carbonates are present in the water (§ 40, *b*).

If a centrifuge is at the disposal of the analyst, the test may be modified so as to include bicarbonates and free carbonic acid. Distilled water is mixed with acetate of calcium and ammonia, the mixture whirled to precipitate traces of calcium carbonate. A sample of water is run into a testing-tube with one or two cub. centim. of the clear mixture, and whirled till the carbonate of calcium is condensed into a thin layer, firmly adhering to the bottom of the tube. This may be hastened by slightly heating the tube before whirling. The water is run off, a little gelatine is spread over the film of precipitate, and after solidification of the jelly some drops of hydrochloric acid.

A large drop is acidulated with acetic acid, evaporated, the residue treated with distilled water, and the decanted solution concentrated to a small drop. This drop is tested for *potassium* with platinum tetrachloride (§ 1, *a*), and afterwards for *magnesium* with sodium phosphate and ammonia (§ 8, *a*).

Another large drop is prepared in the same way, and

the resulting small drop tested for sodium with acetate of uranyl (§ 2, *a*). If the test fails, a little acetate of magnesium must be added (§ **2,** *b*).

b. Two or three small drops of portion (*b*) are concentrated on the same spot in a slide, a little drop of strong nitric acid and a few grains of ammonium molybdate are added and dissolved by cautious heating. Spheroidal grains of *phosphomolybdate* will be found along the border of the drop (§ 52, *b*). Boiling is to be avoided, as it might lead to the precipitation of silicomolybdate (§ 39, *b*).

Another drop of the same liquid is tested for *iron* with potassium ferrocyanide. Afterwards a wire or a thread of glass bent into a ring is put around the drop, an excess of sodium hydroxide is added to the liquid, and the whole covered with a slide that has a drop of hydrochloric acid spread on its underside. It is heated gently till a fine dew begins to appear on the upper slide, and allowed to cool for about half a minute; then the upper slide is removed and the acid mixed with the surrounding drops, a trace of platinum tetrachloride added, and the whole left to spontaneous evaporation. Yellow octahedra indicate the presence of *ammonia* (§ 53, *b*).

c. Two or three drops of portion (*c*) are concentrated on the same spot in a slide. A trace of potassium iodide is added and a few grains of starch. Finally the preparation is touched with a platinum wire dipped in sulphuric acid. The grains of starch will take a tint varying from a grayish-violet to blackish-blue according to the proportion of *nitrite* present.

What is left of the three portions is poured into a small evaporating dish, care being taken to dissolve sulphate of *lead* by warming with a little acetic acid and ammonium acetate. While final concentration on a slide is going on, a little nitrate of copper is added, and after cooling, a drop of a saturated solution of potassium nitrite and a grain of thallous nitrate. Cubic crystals, from dark orange to black, indicate the presence of *lead*.

III. EXAMINATION OF ORES—TRACING OF PRECIOUS METALS

§ 86. Ores with Sulphur, Arsenic, etc.

The examination of ores containing sulphur, arsenic, and antimony presents some peculiarities demanding especial notice.

a. They are at once treated with strong nitric acid, and oxides of *arsenic* and *antimony* isolated by sublimation (§ 69). If *mercury* be present, it will also be detected at this stage of the examination. The residue is treated with nitric acid, which leaves behind sulphate of *lead*, mixed with sulphate of barium and other insoluble impurities.

b. The solution, if treated in the ordinary way, will show an anomalous behaviour, owing to the presence of arsenates.

By treatment with ammonia and ammonium carbonate, *iron* is thrown down as *arsenate, manganese, cobalt, nickel,* and *zinc* as crystalline *double arsenates,* which cannot be distinguished from the double arsenates of *calcium* and *magnesium* (§ 51, *a*), separating out along with them. *Cobalt* and *manganese* may be separated from the rest as insoluble peroxides by heating the nitrates to decomposition and extracting with dilute nitric acid; *nickel* and *copper* as triple nitrites (§ 12, *a*; § 23, *a*). (Further separation by reduction on iron or zinc.) *Zinc* may be detected by heating the double arsenate of ammonium and zinc with sodium carbonate (§ 13, a); by evaporation with hydrochloric acid and sublimation of the chlorides (§ 70); by reduction on a strip of sheet-iron (with a drop of hydrochloric acid), and subsequent extraction with caustic soda and hydrogen peroxide, or sublimation.

For mixed ores containing much arsenic treatment of solution (b) with metallic tin (§ 79, b) is to be preferred.

§ 87. Tracing of Precious Metals

It seems easy work to extract gold and platinum by means of aqua regia, to trace the platinum with cæsium chloride, the gold with stannous chloride, and to crystallise silver chloride from a solution in ammonia. After a few trials with poor ores, only the test for silver is found trustworthy. Sometimes even here a serious difficulty is encountered at the outset, native gold with 50 per cent of silver proving very refractory to solvents.

Small quantities of gold and platinum have a trick of disappearing, while chlorine and the excess of acid are driven off. Probably chloroaurates and chloroplatinates are produced, insoluble in dilute acids. Instead of trying renewed extraction with hydrochloric acid, it is best to avoid the difficulty by employing *extraction in the dry way*.

a. The ore—say an auriferous quartzite with a good deal of pyrites—is crushed to a tolerably fine powder. One or two grammes are put in a small porcelain crucible, moistened with strong nitric acid and heated, the heat being raised slowly to incipient redness. The residue is mixed with twice its volume of borax and fused in a shallow hole on a piece of charcoal with about 0.2 gr. of lead in the midst of the heap. The blowpipe flame must be managed so as to turn the molten mass over several times. By this means small beads of lead produced by fusing with an oxidising flame are incorporated with the main button. The slag is knocked off between double folds of paper.

Refining is done with a small pointed blowpipe flame in a shallow hole, scooped out on a piece of chalk. When the button is reduced to about a quarter of its size, it is left to cool, broken out, and freed from chalk and litharge by tapping and rubbing between paper. The whole operation of melting and refining takes from ten to fifteen minutes. To facilitate disintegration by nitric acid the button is flattened in a steel mortar.

b. Total disintegration may generally be dispensed with. The surface is speedily blackened, and dark heavy particles are thrown off. The solution is then drained off, the lead washed with hot water, rubbing it gently with a platinum wire. This washing is thrown away after separating from it suspended blackish particles of gold. The solution is evaporated with sulphuric acid and *sulphate of silver* extracted from the residue by heating with water. If much silver be present, this solution will, after concentration, deposit clear rhombic octahedra of Ag_2SO_4. It is then tested with potassium bichromate (§ 7, *b*). If poor in silver, it is tested with a trace of ammonium chloride and excess of ammonia (§ 7, *a*).

c. The black particles must be tested for *gold* and *platinum*. Dissolve in a drop of aqua regia, expel the excess, add a small drop of water and a grain of rubidium chloride. If *gold* is predominant, yellow rods (monoclinic) of chloroaurate of rubidium separate out, followed by small brownish-yellow octahedra of *chloroplatinate* of rubidium.

If any considerable amount of platinum be present, these octahedra appear first. If the test for gold should fail, add a grain of thallous nitrate or sulphate (§ 26, *b*). In this way the gold of ores, assaying 30 gr. (1 oz.) in the ton, may easily be traced in 1 gr. of substance. For platinum the limit is to be sought at about a tenth part of this amount.

IV. MICROCHEMICAL EXAMINATION OF ROCKS

In examining samples of rocks the services of microchemical analysis are generally required when determination on crystallographical and physical grounds would be difficult or not quite trustworthy. For investigations of this kind microchemical reactions are combined with crystallographical and physical characteristics of minerals. This circumstance will explain frequent allusions to these characteristics on the following pages.

(1) EXAMINATION OF SLIDES

§ 88. Cleaning

Ready-made slides are a bad material. The cover of thin glass must be removed by heating and pushing it aside. This must be done very cautiously and with a steady motion to avoid tearing the thin slice of rock to small fragments. The uncovered preparation is wiped while yet slightly warm with cotton-wool, saturated with oil of turpentine, then with linen, wetted with spirit. Finally, it is rinsed with water, which must leave a uniform film without any greasy streaks. For fragments of preparations I prefer to push them while hot on to a strip of platinum-foil, and to effect cleaning by heating to low redness. This off-hand method cannot be used when carbonates are to be sought for.

§ 89. Testing for Hard Minerals

When grinding a flake of rock for microscopical examination, begin polishing when a goodly number of translucent specks make their appearance. The main part of the specimen is usually at this stage nearly opaque. In an order for specimens of this description, the thickness may be put at 0.15 to 0.20 mm. for dark-coloured varieties of rock, and at 0.20 to 0.30 mm. for light-coloured ones. For smoothing and giving a partial polish, a slab of cast-iron that has been used some days for grinding is well fitted. Smoothing is performed by rubbing with fine emery and as much water as will make a thin cream. For polishing, nearly all emery is removed by sponging and rubbing with a piece of plate-glass, and copious additions of water are made. If *quartz* be present, isolated bright spots will very soon make their appearance. The order in which the constituent minerals take a polish depends on their hardness and nearly as much on their toughness and the absence of cleavage planes. First of all come *spinels* and *zircon*, closely followed by *tourmaline*, which may, by this device, be detected at the outset and easily distinguished from dark brown hornblende and biotite. After a short time *quartz* becomes bright, together with *cordierite* and *garnets*. *Chalcedony* and *jasper* appear at nearly the same time, while common *opal, olivine, rutile, magnetite, chromite, hematite*, and *pyrite* appear last. Garnets are generally scored with deep fissures; olivine has a dusty appearance; under a lens it looks pitted. Rutile, magnetite, etc.

have a metallic lustre, but not so bright as might be expected from the appearance of their crystals. The *pyroxenes* follow at a marked distance; common augite is easily recognised by fissures and by the rectangular and octagonal shape of transverse sections.

Somewhat later *epidote* appears, and last of all *hornblende*. The varieties of felspar are scarcely brightened on iron; to polish them, a plate of zinc and fine powder of glass or pumice must be employed.

A cursory inspection is made with a pocket-lens; for thorough examination a low power (from 20 to 50) of the microscope is employed, and the specimen conveniently inclined so as to cause a bright reflection.

§ 90. Etching of Polished Specimens

A mirror-like polish may be obtained by rubbing with rouge (ferric oxide) on fine-grained wood or on a slab cast from a mixture of equal parts of shellac and rosin. Little water is used, and the rubbing is continued till the surface becomes dry.

Such polished specimens are sometimes needed for ascertaining in a sample of rock the distribution of fine particles attacked by acids. The specimen is mounted on a broad slide and put face downwards like a cover on a shallow crystallising dish, charged with strong hydrochloric acid and some fragments of anhydrous calcium chloride. An exposure of half an hour suffices for etching olivine; for labradorite an hour must be allowed. On the decomposed minerals minute drops of a saturated solution are found; these are sucked up in capillary tubes for separate examination. If *only* etching is intended, a short treatment with strong hydrochloric acid, assisted by a gentle heat, will answer. After rinsing and drying the specimen is examined under a low power, first lying flat on the stage, afterwards conveniently inclined.

§ 91. Testing for Carbonates

Carbonates diffused through a rock are detected by putting a drop of water on the specimen, covering it and adding a drop of acid in contact with the cover glass, and a narrow strip of filtering-paper on its opposite side. As the water is sucked up, the acid will creep in between the cover glass and the slice of rock, evolving bubbles of carbonic acid, which are speedily arrested by the cover glass. A more elaborate method is described in Part I., § 40, *b*.

§ 92. Staining of Etched Specimens

The *silica* of minerals decomposed by acids may be stained with coal-tar dyes. *Fuchsine*, pointed out by the author in 1871, and once more recommended by him in 1882, and by Haushofer in 1885, has come into general use. Its staining power is very great, and the bluish-red stain comes out with a strong contrast against the surrounding minerals. It must, however, be abandoned, because it is not fast in the presence of light or Canada balsam, and is apt to give accidental films wherever the surface of the specimens is not quite smooth. *Malachite green* is free from these defects, while it surpasses fuchsine in staining power. It is freely soluble even in cold water. *Methylene blue* comes very near it in staining power and solubility, but it is not quite free from the defect of forming accidental films.

Saffranine of commerce (tolusaffranine) is less soluble than the former. It requires hot water and does not give such vivid stains as methylene blue. This defect is compensated for by the absence of accidental films.

Chrysoidine and Bismarck brown are not fitted for staining silica. The stains are faint and do not resist washing with hot water.

§ 93. Methods of Staining

To obtain the full effect of staining, the specimens must be made very thin (from 0.04 to 0.08 mm.), and care ought to be taken in finishing them to efface scratches and remnants of the rough grain produced by grinding.

Basic rocks are etched with strong hydrochloric acid; acid rocks, as quartz-porphyry, syenite, and the majority of granites, require heating with strong sulphuric acid. The etched specimens are rapidly washed, covered with a saturated aqueous solution of the dye-stuff, and left to themselves for a quarter of an hour. A trace of ammonia and occasional heating are useful in promoting the staining effect. After brushing and rinsing in hot water, the specimens are ready to be examined in a wet state, or to be mounted in Canada balsam.

Accidental films are remedied by spreading hydrochloric acid over the specimens and rapidly rinsing them in a large quantity of water.

§ 94. Result of Staining on Minerals

Orthoclase, albite, oligoclase, augite, hornblende, epidote and garnet, when not altered by weathering, are *not* stained. Even sulphuric acid has but little effect on them. Flaws and glass inclusions in the crystals are brought out in strong contrast.

Labradorite, leucite, and olivine are generally stained strongly along the borders of the crystals, faintly in the centre.

Cordierite exhibits the same behaviour as labradorite. By etching and staining it is readily distinguished from quartz; in the latter, only flaws being impregnated.

Of serpentine some varieties are stained without previous etching. Other varieties require etching with hydrochloric

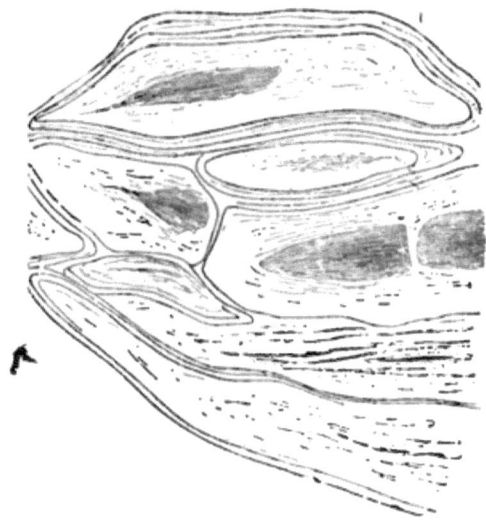

Fig. 77.—Yellow serpentine, Snarum, Norway, stained with malachite green, × ⁹⁄₁.

acid. Sometimes a beautiful network is brought to light on specimens of homogeneous appearance.

Chlorite and mica slates are stained without previous etching, the solution of the dye-stuff penetrating between the flakes. The same remark applies to talc slate and to some varieties of steatite. An aqueous solution of saffranine of moderate strength may be recommended for laying bare the structure of these rocks. Should the wrinkles on the surface be choked with dye-stuff, spread a drop of hydrochloric acid over it and rinse copiously with cold water.

Anorthite, nepheline, elæolite, sodalite, and other minerals which readily yield gelatinous silica are stained vividly, even by saffranine, after previous treatment with hydrochloric acid.

§ 95. Result of Staining on the Ground-mass of Rocks

The matrix of igneous rocks is generally more readily stained than the crystals that have separated out. The most vivid stain is often found where crystals and matrix touch each other, pointing to a greater amount of porosity or weathering along these planes of contact.

On *felsites* and *rhyolites* of homogeneous appearance treatment with strong sulphuric acid and staining with

FIG. 78.—Semi-vitreous felsite, Denver, Colorado, stained with malachite green, × ⁷⁄₁.

malachite green will sometimes reveal perlitic fissures and fluxion bands.

Old basic rocks should be stained twice, having been first treated with hydrochloric acid of sp. gr. 1.12, the second time with stronger acid.

Basalt readily takes a vivid stain when minerals of the nepheline group are predominant in the matrix, also when it is chiefly composed of basic glass.

§ 96. Testing for Phosphoric Acid

For *phosphates* the precipitation as ammonium phosphomolybdate may be adapted so as to yield a localised reaction.

A strong dose of ammonium molybdate and of nitric acid must be added to the ordinary solution, in order to ensure rapid action. Heating is to be avoided, because it might lead to the precipitation of silicomolybdate. The minute crystals of apatite in the slice of rock are wholly dissolved, and their phosphoric acid is immediately precipitated. Specimens treated in this manner have the appearance of being stained. In reality it is not a staining test, but a precipitation, localised in a very narrow area.

§ 97. Testing for Potassium and Aluminium

Two other localised tests are to be mentioned for *potassium* and *aluminium*. They are too tedious for current use.

After being carefully cleaned, the slices of rock are exposed for half an hour in a leaden box to fumes of hydrofluoric acid, given off from a slightly heated mixture of fluorspar and sulphuric acid. After this they are fumigated with sulphuric acid in a platinum crucible. Only a few drops of strong acid are poured in, and a thick layer of asbestos put upon it, to prevent spurting. For a similar reason, the crucible is covered with a piece of asbestos plate. Heating is continued till no more fumes are seen escaping. The specimens are then ready to be tested for *potassium*. This is done with a solution of platinum tetrachloride, to which an equal volume of alcohol has been added. The yellow stains appear immediately; in about half a minute they are fully developed. Washing is done with methylated spirit.

Specimens that are to be tested for aluminium must be immersed in dilute ammonia for about two minutes after the fumigation with sulphuric acid, and put aside to become air-dried. Testing is performed by staining in an aqueous solution of congo red, followed by copious washing with cold water (§ 42, *c*).

§ 98. Disintegration of Rocks

Fractional decomposition of rocks will in many instances render excellent service.

For the effects produced by hydrochloric and sulphuric acids on the more common constituents of rocks, see § 94, § 95.

For further treatment hydrofluosilicic acid (Bořický) or mixtures of ammonium fluoride and hydrochloric acid may be employed; for rapid decomposition hydrofluoric acid, pure or mixed with strong hydrochloric acid.

Hydrofluosilicic acid is a slow solvent, fitted for superficial etching, but not for disintegrating.

Even with mixtures of *ammonium fluoride* and hydrochloric acid the patience of the investigator may be severely taxed by some varieties of basic rocks.

Hydrofluoric acid answers well for the majority of rocks. For refractory varieties its effect is accelerated by strong hydrochloric acid; for acid varieties it is retarded by adding water.

The slice of rock is put on a slide varnished with Canada balsam and covered with a large drop of the solvent. The front lens of the microscope is protected by a small round cover glass, affixed with a drop of water or glycerine.

§ 99. Action of Hydrofluoric Acid on Rocks

a. Quartz is speedily eaten away by hydrofluoric acid.

Felspar is seen to shrink and to become turbid (an opaque white) through incrustation with nearly insoluble fluosilicates and fluoaluminates.

Olivine becomes pitted and furrowed. It is, however, much more slowly attacked by hydrofluoric than by hydrochloric acid.

Garnet, idocrase, cordierite, chlorite, talc, and *mica* are slowly rendered opaque without much shrinkage.

A *vitreous matrix* is sometimes attacked with great violence; when devitrified, it generally resists longer than the crystals of felspar.

Epidote, augite, and *hornblende* stand out in clear relief after the surrounding constituents have become turbid or nearly opaque.

b. The slice is then washed with some drops of water and transferred to a strip of platinum-foil. Here the water is drained off, a drop of strong sulphuric acid is put on, and, by heating, silicon fluoride is expelled. This treatment is repeated till white fumes of sulphuric acid are perceived. Care must be taken not to break the slice by hasty heating, it being partially cemented to the platinum by anhydrous sulphates.

It is loosened by heating with dilute hydrochloric acid

and gently boiled with acidulated water till all soluble matter is removed.

Coarse-grained rocks are apt to crumble in the course of this operation, many fragments showing cleavage planes. Fine-grained varieties throw off a dust that settles speedily into a whitish sediment. It contains minerals of the spinel group, of the andalusite group, tourmaline, epidote, and as main constituents *augite* and *hornblende*. Many crystals are wholly intact, showing all crystallographical details of the species. By judicious application of this treatment, checked by the microscope and eventually assisted by separation in liquids of great density, excellent material for optical and chemical determination may be procured.

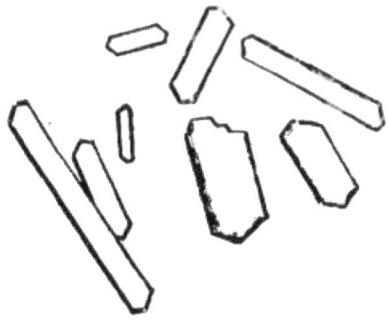

Fig. 79.—Augite from basalt of the Steinsberg, near Suhl, Thuringia, isolated with hydrofluoric acid. ⁿ⸝⁰.

By repeated maceration with hydrofluoric acid such accessory minerals as staurolite, cyanite, tourmaline, zircon, etc., may be completely isolated.

§ 100. Isolating Felspars in Thin Sections

Unfortunately the customary classification of crystalline rocks is based chiefly on the discrimination of the varieties of *felspar*. An expeditious and trustworthy method for isolating them would, therefore, be of great value. As yet nothing remains but picking from coarsely powdered rock or etching on varnished specimens. It is by no means easy to tell if the crystal which is to be etched has been scraped quite clean. Painting the crystal with a solution of dextrine tinged with red ink will be found useful. When dry, the paint is trimmed to size with a needle sharpened to a lancet point, the specimen coated with a thin layer of Canada balsam and stove-dried. After this preparation the crystal may be laid bare by means of the sharpened needle with all desired exactness.

(2) Examination of Powdered Rocks

§ 101. Testing for Hard Minerals

Some small chips of rock are crushed to coarse powder. Grains of quartz, tourmaline, felspar, etc., may be picked out under a good pocket-lens or under a low power of a microscope. This is done with a needle dipped in glycerine, to which the grains adhere on touching. They are transferred to drops of water on a clean slide.

For testing quartz and tourmaline, their superior hardness may be turned to account in the following manner. Put two or three grains on the flat bottom of an agate mortar turned upside down, press a piece of fine-grained wood upon them (the butt-end of a lead pencil), when they will become sufficiently fixed in the wood to bear rubbing forward and backward on a smooth flake of felspar, quartz, or flint. Search for parallel scratches with a pocket-lens after carefully wiping the test-flakes.

Fragments of felspar, when rolled between two pieces of sheet copper, will stick with their sharp edges in the soft metal, and break along cleavage planes.

§ 102. Extraction with Water

In rocks of volcanic origin soluble chlorides and sulphates are sometimes met with. Sodium chloride has also been found in a syenite (Frederiksvärn, Norway) and in quartzite. They are extracted by boiling with water in a platinum spoon. After concentrating, the solution is tested for chlorine with thallous nitrate (§ 60, *a*). The presence of sulphur is generally betrayed by needles of gypsum (§ 21, *a*) without special testing. Aluminium is traced with cæsium chloride (§ 42, *a*). If only traces of chlorides or sulphates are found, the experiments ought to be repeated with an equal quantity of water and no rock, to check the purity of the reagents.

§ 103. Extraction with Hydrochloric Acid

Before macerating with acid, the sample of rock must be ground in an agate mortar to an impalpable powder. Such fine powder is far more easily attacked than the compact substance of a smooth slice of rock.

Orthoclase and cordierite will remain unaltered in slides treated with hydrochloric acid, while the fine powder of the same minerals is sensibly attacked. On the other hand, it is to be borne in mind that streaks and beads of matrix and enclosed needles of other minerals are laid bare in the fine powder. Thus augite from the Firmerich, near Daun, in the southern range of the Eifel, yielded ample indication of sodium and aluminium. A second extraction left the mineral virtually intact, as should be the case with pure augite. The reactions obtained with the first portion of acid were found to have been due to minute specks of nepheline and matrix. Similar cases are of frequent occurrence. They are apt to lead to grave errors if not scrutinised by repeated extraction and by comparison with the results obtained from a careful inspection of slides.

§ 104. Method of Extraction

The operation of extraction is carried out in small platinum spoons of about 10 mm. diameter. The material is transferred into such a spoon by means of a small spatula that will take about 20 mgr. of the fine powder (a hemispherical heap of about 3 mm.) From a small pipette enough hydrochloric acid of 1.12 sp. gr. is run in to cover the powder. It is kept gently boiling for about half a minute, when the powder will be on the point of becoming dry. The volume is then made up with water, heat is applied once more to mix the contents, and a minute allowed for settling.

The solution is transferred to a slide (near a corner) and left to become clear. Filtering is too tedious; turbid solutions are evaporated, the residues gently heated with acidulated water, the clear solutions run off into another corner and there concentrated to the original volume.

Boiling with hydrochloric acid on glass slides should be avoided. If the extraction is performed on a slide, the influence of the glass must be ascertained by evaporating pure hydrochloric acid on it, heating the residue with a few drops of water and a small drop of acid, and testing for alkali metals and calcium. Traces of calcium and sodium, sometimes also of potassium, will generally be found, and these should not be overlooked.

§ 105. Examination of the Solution in Hydrochloric Acid

The solution is divided in three portions—*a, b, c*—by means of a capillary tube.

a. A small drop of platinum solution is added. Traces

of *potassium* will betray themselves when the liquid has nearly disappeared (§ 1, *a*).

b. A little acetate of uranyl and a little acetate of ammonium are dissolved in a drop of dilute acetic acid. This is done close to drop *b*, and the latter concentrated till a saline crust appears around it. At this moment the two drops are made to touch (§ 2, *a*). If the proportion of *sodium* should be too small for a speedy crystallisation of the yellow tetrahedra of the double acetate of sodium and uranyl, some acetate of magnesium is added, which provokes the crystallisation of a triple compound, containing no more than 1.5 per cent of sodium (§ 2, *b*). As acetate of magnesium sometimes contains traces of acetate of sodium, attention is called to the fact that acetate of zinc answers the purpose equally well.

c. Add a small drop of sulphuric acid and heat a few seconds. If no needles appear, heat cautiously till the drop shows a narrow dry seam (§ 21, *a*). For traces of *calcium* wait, as in *a*. When only a film of liquid is left, put in a grain of cæsium chloride and breathe over it (§ 42, *a*). Very small quantities of *aluminium* can be traced in this way. Finally, add ammonium chloride, an excess of liquid ammonia, heat gently and put a grain of sodium phosphate into the warm mixture (§ 8, *a*). For traces of *magnesium* wait about ten minutes.

If the quantity of solution is too small for division, begin as in *a*, go on as in *c* for *calcium*, then test for *sodium* with ammonium fluosilicate (§ 2, *c*), and go on as in *c* for *aluminium* and *magnesium*.

§ 106. Interpretation of Results—Holocrystalline Rocks

If holocrystalline rocks, such as granite, syenite, diorite, etc., are dealt with, the alkali metals and the calcium are derived from *felspar*; they are in this case accompanied by a large quantity of aluminium. Magnesium points to the presence of *biotite*.

If the presence of more than one variety of felspar is to be presumed, stress is to be laid on the proportion of calcium to the alkali metals. *Anorthite* is first attacked; it brings such a quantity of calcium into the solution that sulphuric acid will immediately provoke a copious crystallisation of calcium sulphate. Then follows *labradorite*, yielding much calcium, a little sodium, and often pretty much potassium. After this come *orthoclase* and *microcline*. With them alkali is predominant. Some samples yield almost exclusively potassium (orthoclase from porphyry of Oberried, Black Forest), other samples yield potassium, pretty much sodium, and some calcium (orthoclase from Utö, green microcline from Pikes Peak, Colorado). Last of all come *albite* and *oligoclase*, sodium predominating strongly.

With the four last-named varieties the intensity of the reaction depends chiefly on the degree of weathering. It is a good plan to employ fractional extraction, heating ten seconds with the first, half a minute with the second, a whole minute with the third portion of acid.

§ 107. Hemicrystalline Rocks

With rocks of porphyritic structure the interpretation of the results is complicated by the uncertainty about the composition of the matrix. Yet even here the proportion of calcium to the alkali metals offers a clue as regards the place of the examined rock in the oligoclastic or labradoritic group, provided that comparison be made with the results obtained from standard rocks.

A strong dose of sodium in the first fraction indicates the presence of *nepheline*; crystals of calcium sulphate in the same portion are derived from *nosean* (provided that the extraction with water (§ 101) has been attended to). A strong dose of potassium means that *leucite* is present, possibly concealed in the matrix. A copious precipitate of ammonium-magnesium phosphate points to *basaltic* rocks, even when the microscope does not reveal the presence of *olivine* in considerable proportion. Here, as in many similar instances, comparison with standard rocks is the only way to get out of the confusion, created by hair-splitting classifications. From a chemist's point of view, a certain amount of magnesium, readily extracted by hydrochloric acid, must place a rock of porphyritic structure in the basaltic group, whether plentifully sprinkled with grains of olivine or nearly devoid of them. Let an undoubted basalt, poor in olivine, be adopted as a standard; then two simple tests for magnesium will enable the investigator to decide at once in a quarter of the time that would be spent in comparing half a dozen slides.

§ 108. Fractional Decomposition with Hydrofluoric Acid

When nepheline, olivine, etc., have been destroyed by repeated treatment with hydrochloric acid, fractional decomposition by means of hydrofluoric acid is resorted to. As a rule, little advantage is to be gained from a previous treatment with strong sulphuric acid. By its aid labradorite is wholly decomposed, but at the same time biotite is destroyed and the more acid varieties of mica and felspar sensibly attacked.

If a large batch of samples is to be examined, pure hydrofluoric acid, kept in stoppered tubes of ebonite,[1] may

---

[1] Furnished by Dr. R. Muencke, Berlin, Luisenstr.

be recommended. It saves time and excludes errors in the testing for potassium. For occasional use, ammonium fluoride is to be preferred. Of this solvent about one-half of its volume is added to the remaining rock, hydrochloric acid run in, and a moderate heat applied. The dry residue is moistened with sulphuric acid and heated till no more white fumes are given off. If ammonium fluoride has been employed, the heat must be carried to incipient redness, to make sure that no trace of ammonium remains in the sample. At a low red heat sulphates of aluminium and iron undergo partial decomposition. Dilute sulphuric acid is therefore added and heat applied until the fumes of sulphuric acid become faint. As to dissolving, settling, and testing, see § 103, § 104.

§ 109. Action of Hydrofluoric Acid on the Rock-forming Minerals

A copious reaction of magnesium in this solution is due to *biotite*. Alkali metals and calcium are derived from *felspar*. A little of the remaining powder is examined in a drop of water under a low power, in order to estimate the quantity of solvent needed for destroying the remaining felspar. The second solution generally offers little interest, being derived partly from felspar, partly from pyroxenes and amphiboles. If *cordierite* has been taken for felspar, much *magnesium* and little or no alkali metals will be found in the solution. A great part of the cordierite remains, with augite and hornblende. When all colourless splinters have disappeared, the next fraction will yield decisive tests as regards the nature of the *pyroxenes* or *amphiboles*. Sodium points to *ægirine* and *arfvedsonite*, magnesium to *rhombic* pyroxenes and amphiboles. At the same time the amount of aluminium is diminished; it is often reduced to mere traces.

§ 110. Action of Hydrofluoric Acid on the Accessory Minerals of Rocks

A residue will sometimes remain, resisting obstinately all decomposing agents. If it looks black, or sprinkled with black splinters, search for *tourmaline*, revolving some of the dark splinters over the stage nicol of the microscope. Tourmaline exhibits strong dichroism, ranging from yellow to black. *Staurolite* shows a uniform brown. It is usually perforated with numerous pits and holes, owing to grains of quartz that have been eaten away. *Chromite* is black and opaque; it resists hydrofluoric acid a very long time. Fuse a grain with nitre and sodium carbonate and test with silver nitrate (§ 44, *a*). *Ilmenite* is more speedily attacked; it turns red in hydrofluoric acid, losing titanium (see § 36).

A light-coloured residue may contain andalusite, cyanite, spinel, zircon, or rutile. Splinters of spinel show a very pale red or bottle-green (pleonaste); splinters of zircon a darker brownish-red tint. Both are easily distinguished in polarised light, spinel being isotropic. If picking is to be done, interposition of a thin disc of selenite or mica between the eyepiece and the upper nicol will be found useful.

Rutile, rich in iron, may be confounded with staurolite. Light-coloured rutile looks yellow or orange when reduced to thin splinters. It resists hydrofluoric acid a long time, far longer than ilmenite. Testing for spinel is done by macerating on a varnished slide with ammonium fluoride and hydrochloric acid, covering the drop with a small varnished watch-glass. Spinels will give a crop of octahedra

(§ 42, *b*), but no cubes or hexagons of fluosilicates. The mother liquor contains much magnesium. With regard to zircon and rutile see § 36, § 37, and in this part the section on Rare Elements. Andalusite looks like quartz, cyanite is crushed into thin plates and minute, rod-like fragments. Both show strong polarisation. By hydrofluoric acid they are slowly decomposed, giving a clear solution. From this solution ammonium fluoride precipitates octahedra of ammonium fluoaluminate (§ 42, *b*). The mother liquor contains much silica, but neither alkali metals, calcium, nor magnesium.

V. EXAMINATION OF ALLOYS

(A) GENERAL REMARKS

§ 111. On the Preparation of Specimens

The rough work can generally be done by filing; only the hardest bronzes and hard alloys of iron require grinding. For a rough dressing a revolving grindstone is serviceable, but it will always give concave surfaces which must be planed with emery and water on a slab of iron. For smoothing, emery flour is used on a piece of plate glass; for polishing, tin ash spread on a piece of fine-grained wood. Begin polishing with a few drops of water; go on till the wood becomes dry. By this practice, cleaning with chloroform, alcohol, and ether[1] is rendered unnecessary. For ordinary work extra fine emery paper, glued on a smooth strip of wood, will give a sufficient polish. Small blowholes (gas cavities) are best seen on a mirror-like polish. Streaks and beads of slag show best on a smoothed surface that has been slightly etched.

[1] Prescribed by Wedding, in Stahl. u. Eisen, 1889, p. 264.

§ 112. Examination of Hardness

A want of homogeneity in alloys is often betrayed on smoothed specimens by differences of tint (Cu + Ag, Cu + Sn), or on polished specimens by differences of hardness (hard bronzes and hard alloys of iron). To detect differences of hardness in this way, smoothing and polishing must not be carried too far (see § 89, Testing for Hard Minerals).

For a more detailed examination, needles are employed mounted in handles of the size of a lead pencil. They are used in the way of gravers, digging into the metal, under a low power. The result is sometimes widely different from that which the behaviour of the metal under the file would suggest. This discrepancy must be attributed to the circumstance that resistance to filing is governed quite as much by toughness, elasticity, and close grain as by hardness. The hardness of the following metals and alloys fitted for testing needles is given according to the scale of Mohs (diamond = 10) in universal use with mineralogists :—

Lead = ... 1	Iron (wire tack) = .	. 3.8 to 3.9
Tin = ... 1.7	Sewing needles = .	. 5 to 5.5
Zinc = ... 2.5	Annealed up to the third	
Copper = ... 3	yellow =	. 4
Gun metal = . . . 3.3	Tool steel, annealed up to	
Bronze, 12 per cent tin = . 3.5	the first yellow =	. 6
Bronze, 18 per cent tin = . 3.7	... ...	

§ 113. Colouring of Specimens by Heating

Polished specimens of the alloys of iron and copper may be coloured by heating. The grayish tint of polished *iron* changes to yellow, brownish-red, violet, and blue. This is followed by a pearly sea-green tint, after which the colours recur in the same order as above, every following series becoming fainter than the preceding one. On *copper*, the first change is from light red to dark yellow. This is followed by orange and red of marvellous intensity and beauty, while violet and blue are rather faint and fleeting, being speedily followed by sea-green.

A small admixture of combined *carbon* (up to 1 per cent) will heighten the colours of polished iron; the same holds true for an admixture of *manganese*, which has very strong colouring power. By more than 2 per cent of combined carbon (in white pig and spiegel iron) colouring is reversed; the same is the case with aluminium, silicon, and chromium.

In the alloys of *copper* all admixtures will enfeeble the colour, excepting *manganese*, which seems to exercise an analogous influence to what it has on iron. Alloys of copper and *zinc* are coloured more readily and take more vivid tints than those of copper and *tin;* the alloys of copper and *aluminium* are coloured least of all by heating.

The patterns obtained by heating in homogeneous alloys are nearly allied to those obtained by etching. Etching, indeed, is often combined with colouring. Etching is highly favourable to colouring, hence the possibility of producing striking effects by heating specimens that have undergone a slight etching. But, on the other hand, care must be taken to make all specimens intended for colouring as smooth and clean as possible. Even the combustion products of a Bunsen flame are sufficient to tarnish a cold specimen of steel. Each drop of moisture will leave a trace, which is highly intensified by heating. Colouring should be carried out in an air bath, or in a dish of iron or copper, strongly heated and withdrawn from the flame before the specimen is put in. After a few seconds the heating may be continued, as hot metal is not injured by the combustion gases. On the whole, colouring of alloys by heating is to be regarded as a convenient and elegant, but at the same time somewhat precarious, method of research.

§ 114. Etching of Specimens

Etching is often employed to reveal the structure of metals and alloys. It can be, moreover, utilised for fractional microchemical examination.

The most simple method of etching is furnished by fractional *oxidation* in a *red heat*. Etchings of this kind are frequently found on hardened steel. They have been described by the author of this work when their origin was not fully recognised.[1] On steel they are brought to light by superficial smoothing and polishing. On bronze and brass such etchings by calcination often present themselves when red-hot metal has been quenched in water. Where scales have fallen off etchings are seen, sometimes of singular neatness and beauty. Quenching in weak sulphuric acid will invariably bring them to light; but in this case changes of tint will ensue, betraying a complication which is brought about by the action of the acid and of the metallic solution resulting from it.

Nitric acid has a rapid action, and will generally give a particularly clean and bright surface; but the etchings are apt to become rough, and to lose their neatness, if the action of the solvent is carried a little too far.

---

[1] "Sur la structure de l'acier," Recueil des trav. chimiques dans les Pays-Bas, tom. x., 1891, p. 261.

It is advisable to arrest the action by dipping the specimen in a basin full of water as soon as a brisk effervescence has set in, beginning anew if inspection with the microscope shows the first etching to have been insufficient.

Hydrochloric acid gives excellent results on iron and steel. On gray and white pig-iron, on ferromanganese, ferroaluminium, and ferrochrome, fuming acid is used. Iron and steel must be cleaned carefully after treatment with hydrochloric acid, otherwise the specimens will be speedily deteriorated by rust. On copper alloys hydrochloric acid and aqua regia will produce unsightly films of cuprous chloride. These films may be removed by means of fuming hydrochloric acid or caustic ammonia, but the brilliancy of an etching with nitric acid will never be attained in this way. Both acids are frequently utilised for fractional examination.

Caustic ammonia is very serviceable for etching specimens of copper alloys. It does not give particular brilliancy nor a strong relief fitted for fractional examination; but, on the other hand, the etching is uncommonly smooth and neat, and the action of the solvent very regular and easy of control. No bubbles are evolved, and the progress of the etching may be estimated by the colour of the solvent, which becomes gradually dark blue.

For alloys rich in zinc or tin, *caustic potash* or soda may be employed. For tin, polysulphides of the alkali metals prove to be strong solvents of steady action.

Deeply etched specimens may be subjected to microscopical inspection in the usual position, lying flat on the stage. Polished specimens and such as are slightly etched must be inclined from 10° to 30°, so as to give a bright reflection. A convenient desk for this purpose, with a shallow selvage in front, is easily bent from a strip of thin sheet zinc. To prevent slipping of the slides it is from time to time rubbed with bees'-wax.

§ 115. Specimens for Fractional Examination

The specimens are carefully smoothed to a true plane; they are slightly polished and etched in the ordinary way. The spent solvent is not washed off, but taken up with a capillary tube and put aside for later examination. If, on inspection under a low power, the projecting parts of the relief are found smooth and bright, etch a second time, and so on till the etching has reached a depth of half a millimetre, or till corrosion of the prominent parts sets in. The latter will generally happen with brass and similar alloys, while true bronzes and many alloys of iron allow deep etching. The greatest depth—1.5 mm.—has been attained with ferromanganese and ferrochrome. Finally, samples must be taken of the prominent parts. This is done by rubbing on a piece of agate or flint (bottom of an agate mortar), ground to a true plane, but not smoothed. For examination the metallic film is dissolved in strong nitric acid or aqua regia. Only a few alloys prove refractory to rubbing on agate, such as white pig-iron and ferrochrome with more than 30 per cent of chromium. For such hard alloys a whetstone may be made of powdered corundum worked up into a stiff paste with a hot mixture of equal parts of rosin and shellac. Selected fragments of American corundum are crushed in a steel mortar and ground to flour on a slab of iron with an iron rubber. After repeated macerations with strong hydrochloric acid, boiling with potash lye and thorough washing, it will retain no other impurity but a little titanic acid (rutile) in an insoluble state.

(B) Details of Microchemical Examination

(1) *Iron*

§ 116. Carbon in Iron

a. Combined *carbon* is traced by heating and by etching with nitric acid. Heating will colour steely iron more speedily and more vividly than mild iron (§ 112). On bar iron and steel a drop of nitric acid (sp. gr. 1.2) will produce a blackish stain of carbonaceous matter, mottled or streaked on puddled iron, almost uniform on Bessemer or Siemens-Martin metal. A high percentage of combined carbon protects iron from the corroding action of acids. White pig-iron is etched and tarnished by nitric acid, but not blackened. As the hardening depends chiefly on combined carbon, tests for hardness (§ 111) can often be employed for examining the distribution of combined carbon in hardened specimens.

b. Graphite is readily detected after a strong etching with hydrochloric acid. In gray pig-iron it occurs under the form of curved scales, generally seen edgeways, protruding from narrow rills. In small castings another distribution prevails—namely, minute specks of graphite, composed of minute scales. Under the microscope, graphite shows a light gray and sub-metallic lustre. From ferrosilicon it is easily distinguished by digging into it with a needle.

§ 117. Silicon in Iron

a. Silicon may be separated from iron by sublimation

with ammonium fluoride. The solution in nitric acid is heated with sulphuric acid in a platinum spoon till the residue is nearly dry; then ammonium fluoride is added with a small drop of water. Another platinum spoon, cooled with a drop of water, is used for cover during the sublimation. The sublimate is tested with sodium chloride on a varnished slide (§ 39, a).

b. Testing with ammonium molybdate and rubidium chloride (§ 39, b) demands great care in order to avoid errors. Dissolve the sample of iron in nitric acid, add ammonium molybdate and a little ammonium carbonate. Heat gently, to ascertain if phosphoric acid be present. If this is the case some minutes are allowed for settling, the clear liquid is run off into the next corner of the slide, mixed with some drops of water and a drop of nitric acid, and heated to ebullition. If much silicon is present, yellow grains of ammonium silicomolybdate will separate out; if this does not happen, rubidium chloride is added. An excess of rubidium chloride will produce orange prisms of a double chloride of iron and rubidium, dissolving freely in water.

§ 118. Phosphorus in Iron

Testing for *phosphoric* acid is done with ammonium molybdate at ordinary temperature or at a gentle heat (§ 52, b; § 117, b). A complication may arise if tungsten or molybdenum be present. In this case a granular precipitate of phosphotungstate or phosphomolybdate will come down after addition of a salt of potassium or ammonium.

§ 119. Sulphur in Iron

Sulphur is not easily oxidised to sulphuric acid. Even

by aqua regia sulphuretted hydrogen may be evolved. A solution of bromine in hydrochloric acid is the safest solvent. The solution is twice evaporated with nitric acid. The residue is stirred up with a drop of water and a little acetate of calcium; it is then dried, and the basic nitrate of iron decomposed by heating. The crust of oxide is treated with hot water, the clear solution run off into the next corner, and concentrated till crystals of gypsum separate out (§ 54, *a*). For very small quantities of sulphur the acid solution is evaporated, a drop of water and excess of ammonia are added, and run off after heating to the next corner. The ammoniacal liquid is evaporated with a trace of sodium carbonate; ammonium salts are expelled by heat, and testing done with dilute acetic acid and a little calcium acetate (§ 54, *a*). If ferric oxide separates out, evaporate, heat with water, and concentrate the clear solution. For traces of sulphur, lead acetate (§ 54, *c*) is employed instead of calcium acetate.

§ 120. Manganese in Iron

a. Manganese is easily detected in the dry way. The sample is dissolved in nitric acid, the solution is evaporated, and the residue fused on a platinum wire with sodium carbonate. With 0.5 per cent of manganese an unequivocal green tint is obtained. A great part of the iron may be eliminated by employing the method of precipitation pointed out by Hampe (§ 10, *c*). The solution in nitric acid is heated to ebullition with potassium chlorate till a dark brown precipitate of manganese peroxide is formed. This is washed with dilute nitric acid and tested as above.

b. If preference be given to the wet way, the precipitate

is dried and dissolved in hydrochloric acid. If the solution looks decidedly yellow, it must be evaporated with nitric acid, and precipitation with potassium chlorate repeated. Testing is done in a neutral solution with oxalic acid (§ 10, *a*). If no crystals appear, the delicacy of the test may be considerably increased by adding an excess of ammonia, which will produce colourless spear points and fringed rods, often united so as to form spoked aggregates.

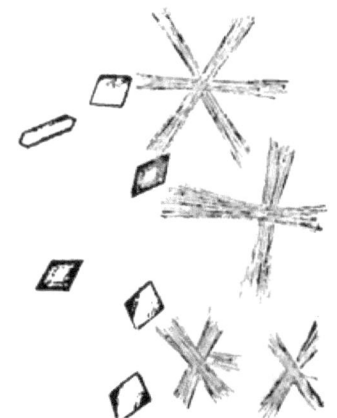

FIG. 80.—Manganese oxalate, treated with ammonia, ×130.

§ 121. Chromium in Iron

Rich ferrochromes are difficult to dissolve in nitric acid. They are best decomposed in the dry way. For chromium-steel and poor ferrochromes (up to 20 per cent of Cr) the wet way has several advantages.

a. First Method.—The metal is reduced to powder, this is partially oxidised with nitric acid, dried and fused on platinum with sodium carbonate and nitre. Any excess of reagents must be avoided. The caked mass is lixiviated with water, the clear solution concentrated, acidulated with acetic acid, and precipitated with silver nitrate (§ 44, *a*). If the sample is very small, as in fractional examinations, a little sulphuric acid is added with the acetic acid. In this case great rhombic crystals are formed, coloured from light orange to fiery red. They contain Ag_2SO_4 and Ag_2CrO_4 in varying proportions.

b. Second Method.—Dissolve in an excess of nitric acid, add potassium chlorate and heat to ebullition. Go on adding small quantities of chlorate till the colour of the solution has become a pure orange; then add acetate of lead, dilute with water and heat, to dissolve nitrate and chloride of lead. The chromate is precipitated as a powdery yellow film, adhering to the glass. It is twice washed with hot water and made to crystallise as basic chromate by putting a minute grain of caustic potash upon the wet film. Part of the chromate is dissolved; crystallisation ensues, where the saturated solution encounters grains of undissolved chromate. The crystals of the basic chromate are short, fringed rods, grouped in sheafs (40 to 80 μ). Their colour varies from bright reddish-orange to brick-red.

FIG. 81.—Basic chromate of lead, ×90 : 1.

§ 122. Tungsten in Iron

Tungsten is oxidised by nitric acid to tungstic acid, but the action of the solvent is rather slow. Though not so refractory to solvents as ferrochrome, ferrotungsten may be etched with aqua regia to a depth of 0.5 mm. Tungstic acid is precipitated from the acid solution as ammonium phosphotungstate. The solution is strongly acidulated with nitric acid, a small drop of ammonia is added, and a very small quantity of sodium phosphate. A gentle heat promotes settling of the phosphotungstate (§ 58, *b*). It is washed twice with a drop of water and changed to thallous tungstate (§ 58, *c*) by adding first a little caustic potash and after a few seconds a grain of thallous nitrate. One-half per cent of tungsten is easily detected.

§ 123. Aluminium in Iron

a. Aluminium.—Three per cent of aluminium are detected without special precautions. Dissolve in strong hydrochloric acid or in aqua regia, evaporate with a small drop of sulphuric acid, place a small drop of water close to the residue, add a grain of cæsium chloride, and let the aqueous solution come in contact with the residue. Six-sided yellow plates of Cs_3FeCl_6 are formed besides colourless octahedra of cæsium alum (§ 42, *a*). With some care and patience this test will yield satisfactory results with 1.5 per cent of Al. It is necessary to add a small drop of water, when some alum crystals are seen. It will dissolve the iron compound, doing little injury to the alum crystals, which will develop into splendid specimens.

b. For small quantities of aluminium (as low as 0.5 per cent), evaporate the solution in aqua regia with a little sulphuric acid and potassium sulphate, wash with alcohol, moisten the residue by the breath, allow to dry, and wash once more with alcohol. In this way a great part of the iron is removed as chloride. Potassium alum remains undissolved with a little ferric sulphate and ferric chloride. Dissolve in a drop of water, concentrate and precipitate with cæsium chloride, adding a trace of sulphuric acid.

§ 124. Nickel and Copper in Iron

a. Nickel is traced as triple nitrite of nickel, lead, and potassium (§ 12, *a*). The solution in nitric acid is evaporated, the residue dissolved in dilute acetic acid. Add a little sodium acetate, potassium nitrite, and a grain of lead acetate. This test is very delicate.

b. Copper is detected as triple nitrite (§ 23, *a*), operating in the same way as for nickel. If a grain of thallous nitrate be added at the end, it will considerably increase the delicacy of the test.

(2) *Copper and its Alloys*

§ 125. Cuprous Oxide in Copper

Cuprous oxide, the most common impurity of copper, is easily detected by etching polished specimens with ammonia. Cuprous oxide dissolves more speedily than metallic copper. If any considerable amount of cuprous oxide be present, the etched surface will exhibit an irregular pattern of rudimentary cubic crystals, separated by deep narrow rills, looking like cracks, where cuprous oxide had separated out on the joints of the crystals.

§ 126. Sulphur, Phosphorus, and Arsenic in Copper

Sulphur and *phosphorus* may be detected in copper by means of the same tests as in iron.

Arsenic.—To detect this impurity, dissolve in nitric acid, heat with a grain of potassium chlorate, add ammonium molybdate, and wait some minutes for precipitation of phosphomolybdate. If no granular precipitate is formed, apply a gentle heat, to precipitate the arsenomolybdate. After a slight washing the precipitate may be dissolved in ammonia, and the presence of arsenic acid fully established by precipitating characteristic crystals of ammonium-zinc arsenate (§ 51, *a*).

§ 127. Antimony, Bismuth, and Lead in Copper

Small quantities of *antimony* and *bismuth* are difficult to detect. See Antifriction Metals, § 143.

Lead is easily traced. The dry residue of a solution in nitric acid is dissolved in dilute acetic acid, a little sodium acetate is added, and after this potassium nitrite. If after a minute no cubic crystals (§ 23, *a*) are seen, the delicacy of the test may be trebled by addition of a grain of thallous nitrate. If the admixture of lead rises above 1.5 per cent, it will produce a grayish network on etched specimens, the grayish threads standing out in relief. A similar effect is produced by 2 per cent of zinc, by 1 per cent of tin, and by 0.5 per cent of silver.

§ 128. Alloys of Copper—Copper and Tin, Bronze

From 2 to 6 per cent of tin will produce a network of yellowish threads in a reddish metal. Bronzes,

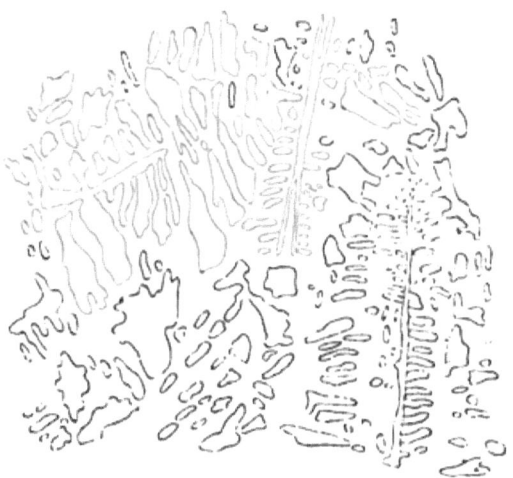

FIG. 82.—Gun bronze, 10 per cent of tin, × 50 : 1.

containing from 8 to 15 per cent of tin, show cruciform and dendritic groups of cubic crystals, from orange to dark yellow, in a homogeneous matrix of a lighter yellow. In bronzes with 15 to 20 per cent of tin the matrix is nearly white, the crystals light yellow, very small, arranged on lines that cross each other at right angles. The matrix is harder than the crystals; when white, its hardness rises above 4 (copper = 3); at the same time it shows more resistance to solvents. Caustic ammonia is a very good solvent for etching bronze. It chiefly dissolves the copper; at a more advanced stage of the etching, however, tin is also dissolved as stannous oxide, probably by a secondary action of the ammoniacal solution of cupric oxide.

§ 129. Impurities in Bronze

a. The presence of *tin* is established by dissolving the alloy in nitric acid and heating to ebullition. The tin is thrown down as metastannic acid, which is slightly washed, heated with hydrochloric acid, and tested with cæsium chloride (§ 35, *a*). If much copper be present, the colourless octahedra of the stannic compound are accompanied by brownish prisms of a double chloride of copper and cæsium, freely soluble in water.

b. The small quantity of *phosphorus* (generally less than 1 per cent) present in phosphor-bronze combines with tin; such bronze is treated with nitric acid. Stannic phosphate is produced, a very stable compound, almost insoluble in nitric acid. Boiling with water and sodium carbonate will decompose it. From the solution, ammonium-magnesium phosphate may be precipitated (§ 52, *a*) with ammonium chloride and magnesium acetate; the residue may be tested for tin as above (§ 129, *a*).

c. Small quantities of *lead* are frequently found in bronze. The lead is detected as triple nitrite (§ 127).

Zinc is often present in great quantity. See Brass and Statuary Bronze, § 130, § 132.

§ 130. Copper and Zinc, Common Brass

These alloys are more homogeneous than true bronzes. Spots of white alloy are never found. By etching, a yellow lattice-work on a reddish ground is laid bare in alloys containing from 3 to 12 per cent of zinc, or a yellow weft, filled in with darker yellow (alloys with 20 to 50 per cent of zinc). Caustic ammonia will give good etchings; its only drawback is a broad dark seam of precipitated copper. This is very characteristic of brass, but at the same time it looks somewhat unworkmanlike. As it is difficult to remove, it is avoided by immersing the specimens in ammonia instead of pouring on the solvent. Heating, followed by quenching in weak sulphuric acid, will give good characteristic etchings: on *bronze*, a red field with whitish threads; on *brass*, an orange field with yellow threads. Strong nitric acid will produce a marked relief on *bronze*; on *brass* the relief will be insignificant, the colour of the etched specimen being a brilliant gold yellow.

The *hardness* of brass is inferior to that of bronze. It remains stationary between 3.1 and 3.2, from 8 per cent up to 50 per cent of zinc. No sensible difference of hardness is to be found between crystals and matrix.

§ 131. Impurities in Brass

To establish the presence of *zinc*, dissolve in nitric acid, evaporate, if possible, on the sample of brass, add water and evaporate once more. Much copper is thrown down, while zinc is taken up. Dissolve in water, evaporate on glass or platinum, decompose the nitrates by heating, evaporate with a small drop of water and a little caustic soda, and wash with cold water. Zinc is dissolved, cupric oxide is left behind. The solution is concentrated at a gentle heat, some grains of ammonium carbonate are added, and time is allowed for volatilisation of ammonia at ordinary temperature. If the percentage of zinc is low, the sample must be left to become quite dry, when a drop of water will lay bare sharp colourless tetrahedra of the binary carbonate of sodium and zinc (§ 13, *a*).

Lead is detected as triple nitrite.

Iron is present in Delta metal to the amount of from 0.5 to 2.0 per cent. It is readily detected as Prussian blue (§ 43, *a*) after precipitating the copper on a piece of pure laminated zinc.

§ 132. Statuary Bronze

Statuary bronze is generally a reddish brass, alloyed with tin and lead in varying proportions. If filings or scrapings can be procured the examination is easy. Treatment with nitric acid will leave *metastannic* acid behind (§ 129); the solution is tested for *zinc* and *lead* in the same way as a solution of common brass. If filing or scraping be not permissible, rub a smooth part of the object with a wet piece of linen, slightly powdered with emery flour. All linen, not impregnated with metal, is cut away; the rest is extracted at a gentle heat with hydrochloric acid, to which a drop of nitric acid has been added. The solution is twice evaporated with nitric acid. By this treatment organic impurities are destroyed, and *tin* is thrown down as metastannic acid, which is left behind, when the residue is treated with weak nitric acid. This solution is tested for *zinc* and *lead* as indicated above.

§ 133. Copper and Aluminium, Aluminium Bronze and Aluminium Brass

Aluminium bronze (copper and aluminium) can be distinguished from aluminium brass (copper, zinc, and aluminium) by pouring caustic ammonia on smoothed specimens, when aluminium brass will speedily show a tarnished dark seam. Another means of discrimination is furnished by the structure. Aluminium brass exhibits the fine rectangular latticework of common brass with, however, a more marked relief. Aluminium bronze of low grade (2 per cent to 6 per cent of Al) will show the same network as common low-grade bronze; with from 8 per cent of aluminium upwards a felted structure asserts itself, becoming very marked between 10 and 12 per cent of aluminium. The structure must be laid bare by etching with ammonia or with nitric acid; heating is of no avail either for etching or for staining. In several treatises on Metallurgy it is stated that aluminium bronze with 10 per cent of aluminium is an homogeneous compound (Cu_4Al), and further, that its hardness is superior to that of copper-tin bronze. These statements are erroneous; the second being probably deduced from experiments with an alloy rich in silicon. Both aluminium bronze and aluminium brass differ very little from common brass in hardness, but are far superior to it in tenacity and smoothness under the file. Aluminium bronze of 10 per cent Al is deeply scratched by gun-metal of low grade (8 per cent of tin).

§ 134. Copper with small Percentage of Aluminium

Direct testing for *aluminium*, with sulphuric acid and cæsium chloride (§ 123, *a*), is admissible to as low a figure as 3 per cent of aluminium. For more delicate work copper and zinc are precipitated from a neutral solution with oxalic acid. To make the precipitation complete the liquid must be evaporated, the residue extracted with water, and this treatment must be repeated a second time. The last solution is evaporated with sulphuric acid and the residue is heated till white fumes of sulphuric acid are evolved. It is then tested in the ordinary way (§ 42, *a*).

§ 135. Copper and Silicon—Silicon Bronze and Cowles's Bronzes

The influence of silicon on the colour and structure of copper is four times greater than that of tin; it is twice as great as its influence on the hardness. The structure of copper, alloyed with 1 per cent of silicon, is similar to that of common bronze with 4 per cent of tin. Above 5 per cent the hardening influence of silicon is very marked; the metal becomes almost white, and curious gray needles and plates make their appearance. In a 10 per cent cuprosilicon they are an essential constituent. In colour and lustre they resemble graphite, in hardness they surpass felspar. There is but one solvent that has any effect on them—a mixture of nitric and hydrofluoric acids. They consist of crystallised silicon and a trace of copper.

§ 136. Effect of Silicon on Copper Alloys

The colour of ordinary bronze is made whiter by an admixture of cuprosilicon, while its hardness and its malleability are but slightly influenced. An alloy containing 6 per cent of tin and 2 per cent of silicon resembles bell-metal with 20 per cent of tin in colour and in the closeness of its crystalline structure. It allows cold forging, in hardness it stands close to gun-metal with 8 per cent of tin.

§ 137. Testing of Copper Alloys containing Silicon

The examination of these alloys may be performed in the following way. By treatment with nitric acid *tin* is converted into insoluble metastannic acid; silica is dissolved along with copper. It is precipitated as sodium fluosilicate (§ 39, *a*) by adding ammonium fluoride and sodium chloride on a varnished slide. If the alloy is rich in tin and poor in silicon, some stannic oxide, suspended in the solution, might occasion an error, through a faint resemblance between the fluostannate and the fluosilicate. In this case absorb the mother liquor with a rolled piece of filtering-paper, wash with a single drop of water, transfer the sodium salts in a drop of water to an ordinary slide and heat with barium acetate, thereby producing the characteristic barium fluosilicate, which cannot be confounded with the analogous fluostannate.

§ 138. Copper with Silicon and Aluminium

Cowles's bronzes (copper, alloyed with aluminium and silicon) are harder than the corresponding alloys of copper and aluminium. An alloy containing 10 per cent Al and 2 per cent Si is harder than iron; with 10 per cent Al and 3 per cent Si the metal is grayish, somewhat brittle; its hardness (5.0) is equal to that of ordinary hardened steel; at many points it is equal to that of hardened tool steel. On these points gray needles are perceived, probably of crystallised silicon. The best solvent for these remarkable alloys is aqua regia. In the solution aluminium can be traced with sulphuric acid and cæsium chloride, but then sublimation with ammonium fluoride becomes necessary for isolating the silicon. A better method is to precipitate the main bulk of the *aluminium* as fluoaluminate of ammonium (§ 42, *b*) and afterwards to trace the silicon in the same drop as fluosilicate of sodium (§ 39, *a*). This may be done in the following way:—Concentrate the solution in aqua regia to a pasty mass. This is transferred to a drop of water on a varnished slide, in which an excess of ammonium fluoride and some ammonium acetate have been dissolved. When the octahedra of ammonium fluoaluminate have attained their full size, the mother liquor is run off into the next corner. A little ammonium fluoride and ammonium acetate is added, to make sure that no more fluoaluminate will separate out. If only a few small crystals are formed, sodium chloride may be added. It will precipitate a granular mass of cryolite, and at a greater distance hexagonal plates and stars of sodium fluosilicate will separate out. If too much cryolite is thrown down, run the mother liquor off and let it dry up; then a drop of water will reveal the characteristic crystals of sodium fluosilicate.

§ 139. Copper and Manganese—Manganese Bronze and Manganese Brass

The effect of manganese on copper is analogous to that of nickel. Ten per cent of manganese makes the colour a faint red; with 20 per cent the alloy is grayish; with 30 per cent it is nearly white. A yellow alloy has not been obtained. All alloys up to 30 per cent of manganese are malleable. Their hardness varies from 3.0 to 3.2. *Manganese brass* of low grade does not differ essentially from common brass either in hardness or in structure; with equal parts of manganese and zinc (15 per cent of each) it has the colour, structure, and hardness of German silver. *Manganese bronze* is more close-grained, at the same time it partakes of the hardness of the copper-tin alloys. An alloy containing 10 per cent of manganese and 5 per cent of tin is almost white; its hardness is on a level with that of gun-metal (8 per cent Sn).

To examine an alloy of this kind, precipitate *stannic oxide* by heating with nitric acid, add potassium chlorate to the solution and boil, to precipitate *manganese peroxide* (§ 120). Wash with dilute nitric acid, and test the washing for zinc in the same way as a solution of brass (§ 131).

§ 140. Copper and Nickel—Nickeline, Manganine, German Silver

All these alloys are light gray with a yellowish tinge; their hardness scarcely exceeds that of brass (3.0 to 3.2). The structure of nickeline and of German silver is a coarse network; manganine is granular with very close grain. In examining these alloys, first try to precipitate *manganese peroxide* (§ 120). Then metallic *copper* is precipitated on a bright piece of sheet iron. The reduction is promoted by a gentle heat, and by addition of a little sulphuric acid. To be assured that no copper remains in the solution, draw a part of it aside, add a trace of sulphuric acid and see if a red film is deposited in the course of a minute. If the iron is not reddened, transfer the solution to a slide, add acetic acid and sodium acetate, and test for *nickel* with potassium nitrite and lead acetate (§ 12, *a*). The mother liquor is evaporated with an excess of ammonia, the residue is washed and tested for *zinc* with caustic soda and ammonium carbonate (§ 131).

(3) *Alloys of Lead, Tin, and other Metals*

§ 141. Lead, Tin, and Antimony

Alloys of lead and tin, in widely varying proportions, are in current use. Alloys of lead and antimony (hard lead) are the chief material for printers' type. Pewter and Britannia metal have tin and antimony for main constituents, with varying quantities of lead, and sometimes with small quantities of copper and bismuth. Alloys of a similar complicated nature are also employed for casting small printers' type.

a. An admixture of *tin* in lead is detected in the same way as in alloys of copper (§ 129). Tracing *lead* in tin is of far more importance. Wrought metal, foil, tubing, etc., must be treated with nitric acid. The solution contains lead and other impurities (copper, iron, zinc). Drive the excess of nitric acid off, add acetic acid, sodium acetate, potassium nitrite, and if the utmost delicacy is desired, a grain of thallous nitrate. If dark brown or black cubes appear, the presence of *lead* and *copper* is established. Otherwise, one-half of the sample is tested for lead with sulphate or nitrate of copper, the other half for copper with acetate of lead (0.05 per cent of lead is detected without any difficulty).

b. In cast metal 0.5 per cent of lead can be detected by etching with a solution of iodine in hydriodic acid prepared by dissolving potassium iodide in hydrochloric acid, and adding a trace of potassium nitrite. While the solvent acts on the tin nitrite is added to keep up a pale sherry colour. After a minute or two a yellow network will appear where an alloy, *rich in lead*, has accumulated on the outside of the tin crystals. From this experiment it is evident that legal restriction of the amount of lead to 1 per cent will not afford absolute security. On objects finished with the burnishing steel the test will not succeed unless the burnished surface be taken off by rubbing with fine emery paper.

c. Hard lead is treated with nitric acid. The greater part of the *antimony* remains undissolved as antimonous oxide. For testing see § 49, *a*. Alloys containing *tin* and *antimony* leave a mixture of metastannic acid and antimonous oxide soluble in hydrochloric acid. *Tin* may always be traced with cæsium chloride (§ 35, *a*). Any considerable admixture of *antimony* is betrayed by the same reagent (§ 49, *a*). For small quantities of antimony and copper the treatment must be modified. The solution in nitric acid must be concentrated till it is on the point of drying up. Hot water will dissolve nitrates of lead and copper, leaving behind metastannic acid, antimonous oxide, antimonates, and basic nitrates of copper and bismuth. This residue may be dissolved in hydrochloric acid to be tested for *tin*. When this has been done, a separation can be effected by heating gently on a bright piece of metallic tin. A black or reddish coating of metallic antimony, copper, and bismuth is deposited. For more detail see § 143, *d*.

d. Arsenic is detected in lead by dissolving in nitric acid, heating with potassium chlorate, and precipitating the arsenic acid with ammonium molybdate (§ 51, *b*). The yellow precipitate may be washed, dissolved in ammonia, and tested for arsenic acid with magnesium acetate. The test with calcium acetate (§ 51, *a*) is rendered difficult by a copious precipitate of calcium molybdate.

Alloys of *tin* and *arsenic* are difficult to examine. Treatment with nitric acid and potassium chlorate will produce a white precipitate composed of metastannic acid and stannic arsenate ($Sn_2As_2O_9 + 10H_2O$). The latter compound is almost insoluble in nitric acid; by heating with dilute nitric acid and ammonium molybdate it is slowly decomposed. If the white precipitate is calcined with sodium carbonate, subsequent treatment with water will leave much stannic oxide undissolved. The solution may be tested with nitric acid and ammonium molybdate, but the result is rather unsatisfactory. A better result is obtained by the following method:—The white precipitate is dried, moistened with nitric acid and dried again, to destroy a remnant of stannic chloride. Boiling with water and sodium carbonate will then dissolve sodium arsenate and a little stannate. Stannic oxide is precipitated by heating with ammonium chloride. The clear solution is then tested for arsenic acid with zinc acetate and ammonia. Ammonium-zinc arsenate exhibits the same forms as the corresponding compound of calcium, which is not available in this case, by reason of the great quantity of soluble carbonate. This modification of the test is, however, as conclusive for arsenic as the test with a calcium compound. If the proportion of arsenic is very low, a great part of the tin may be removed by treatment with fuming hydrochloric acid, to which now and then a drop of nitric acid is added. A final heating will prevent any loss, precipitating all arsenic that may have been dissolved.

§ 142. Alloys for Clichés

These are composed of *tin, lead*, and *bismuth* ; Wood's and Lipowitz's metal of tin, lead, bismuth, and *cadmium*.

Observation of the melting-point may afford a clue to the composition. Heat a chip of the alloy on a strip of sheet iron close to a small piece of plumber's solder; it will melt before the solder if bismuth be present. Heat in water, acidulated with a drop of hydrochloric acid; if the alloy does not melt, the presence of cadmium is highly improbable. For a full examination, dissolve a sample in nitric acid, evaporate till the residue is nearly dry, heat with a large drop of water and decant the solution containing lead, cadmium, and a trace of bismuth. The residue is treated with nitric acid, the excess of acid is driven off, and the presence of *bismuth* is established by testing with acid potassium oxalate or potassium sulphate (§ 48, *a, c*). Any residue is dissolved in hydrochloric acid and tested for *tin*. In the solution containing lead and cadmium, *lead* is detected by adding a trace of potassium iodide (§ 22, *b*). It is then evaporated with sulphuric acid, cadmium sulphate is redissolved in water, and *cadmium oxalate* is thrown down by adding potassium oxalate (§ 14, *b*). If the precipitate shows a suspiciously microcrystalline appearance pointing to *zinc*, heat it with sulphuric acid, add an excess of caustic ammonia, and effect the separation of cadmium and zinc with sodium bicarbonate (§ 13, *a*). The presence of cadmium may be established by exposing the crystals to a current of sulphuretted hydrogen. They will be stained a bright yellow if cadmium be present.

§ 143. Alloys for Bearings, Antifriction Metals

Some analyses will give an idea of the varying composition of alloys for bearings and cushions:—

	Sn	Sb	Cu	Pb	Zn
Prussian railways	{ 91	6	3	...	...
	{ 74	15	11		...
Bavarian railways	90	8	2	...	
Magnolia metal	5.9	16.3	...	77.8	...
Fenton's metal	...	14.5	5.5	...	80

a. Zinc is detected in the same way as in brass (§ 131). If *lead* be present this is immediately precipitated from the alkaline solution when ammonium carbonate is added. The liquid is then decanted and left to deposit the tetrahedra characteristic of zinc. From Fenton's metal dilute sulphuric acid dissolves zinc with a trace of copper. The solution may immediately be tested for zinc with sodium bicarbonate (§ 13, *a*).

Lead and *copper* are detected in a small drop of the original solution in nitric acid by means of potassium nitrite (§ 141, *a*).

b. Bismuth may be separated from other metals by sublimation of its chloride. The solution in nitric acid is twice evaporated, the second time with dilute hydrochloric acid. Hot water will remove the greater part of the lead. The residue is treated according to § 70. If the heat has not been carried beyond a low red, *bismuth* will so far predominate that its presence in the sublimate may be easily established by testing with acid potassium oxalate or potassium sulphate.

c. Antimony may be traced in the sublimate, but its presence cannot be fully established unless its proportion in the composition of the alloy rises to about 20 per cent. A conclusive test is found in the precipitation of sodium antimonate, which will give good results with alloys containing 3 per cent of antimony. The residue of oxides and basic nitrates is fused with nitre (§ 49, *d*); the fused mass is heated with water and a little caustic potash. The solution is tested with sodium chloride and alcohol.

d. If the alloy is chiefly composed of tin and lead, treat with nitric acid and evaporate in a porcelain crucible. Moisten with hydrochloric acid; evaporate once more; heat with a large quantity of water. Copper, tin, and nearly all the lead will be dissolved. Treat the residue with hydrochloric acid, and precipitate metallic antimony, bismuth, and copper by heating the solution on a piece of tin. The dark coating of powdery metal is detached by a rapid heating with strong hydrochloric acid. It is washed, oxidised with nitric acid, and fused with nitre for *antimony*. *Bismuth* is traced in the residue from the solution of potassium antimonate.

(4) *Alloys of the Precious Metals*

§ 144. General Properties

All alloys of precious metals have a marked tendency to crystallise, splitting up into alloys of different fusibility. Thus, on etched surfaces, 3 per cent of *gold* in silver is betrayed by yellow threads on a white field. Two per cent of gold may be detected on specimens of its alloys with copper. The same limit is found for alloys of *platinum* with copper or zinc. With *silver*, liquation goes much further. In *zinc* 0.5 per cent of silver will start a copious crystallisation of the compound AgZn, easy to detect on smoothed surfaces by etching with caustic alkali or with sulphuric acid. In *copper* 0.1 per cent of silver will produce fragmentary whitish threads ; 0.5 per cent will produce an unbroken net in buttons of 1.5 gr. One per cent may without difficulty be detected on flattened buttons of 0.3 gr., fused on charcoal in the blowpipe flame. As this test is very convenient,

FIG. 83.—Copper with 1 per cent of silver, etched with nitric acid, ×40 : 1.

some details may be useful. A sample of 0.2 to 0.4 gr. is fused on charcoal. If a reducing flame is employed borax is rendered unnecessary. When the metal is fused to a round button, blowing must be diminished, to let solidification take place gradually and in a reducing atmosphere. Spurting can generally be prevented by this practice. If it should occur nothing remains but remelting with a little borax. The buttons are flattened in a steel mortar to discs of 5 mm. diameter; these are dressed on one side with a smoothing file, and finished on a fine whetstone. Rubbing the stone with a very little petroleum will be found helpful in polishing small specimens. Etching is done with strong nitric acid. Put the specimen on a dry slide, touch with a glass rod of the same diameter dipped into the acid, drop the specimen into a basin of water when the effervescence has spread over the whole surface; after rubbing with linen it is ready for inspection. Crystallisation will sometimes be found to have started from several points. In this case the network will be seen to be interspersed with radiated figures. If these irregularities could be avoided, such etchings might be utilised for rapidly estimating small quantities of silver.

Colouring by heating has a striking effect on poor alloys of gold or silver with copper. If the amount of precious metal exceeds 20 per cent, the colour becomes faint.

§ 145. Gold Alloys

a. Alloys of *gold* and *silver* are easy to examine if the proportion of gold is below 30 per cent. *Silver* is extracted with nitric acid (for testing, see § 7, *a, b*); afterwards *gold* is dissolved in aqua regia and tested with stannous chloride or with thallous nitrate (§ 26, *a, b*). Samples, composed of equal parts of gold and silver (some varieties of native gold) are refractory to solvents. They must be alloyed with thrice their weight of lead in the blowpipe flame, or with cadmium, melted under a layer of potassium cyanide in a porcelain crucible. Silver is dissolved with the base metal in nitric acid. If cadmium be employed, the solution should be immediately tested with potassium bichromate. If lead be used, evaporate, wash with a drop of cold water, which will leave a great part of the nitrate of lead undissolved, and precipitate the rest by evaporation with sulphuric acid. Redissolve in hot water, and add potassium bichromate. Alloys, in which gold predominates, are attacked by aqua regia; gold is dissolved, chloride of silver remains as an adherent coating, sometimes stained yellow by a chloroaurate of silver. It is dissolved in ammonia, and brought to the test by volatilisation of the solvent (§ 7, *a*).

b. Alloys of *gold* and *copper* are heated with nitric acid. If they are low in gold, the latter is disintegrated to a dark brown powder. With flattened buttons a short treatment generally suffices; the dark coating is brushed off with a platinum wire. The dark particles are rapidly washed, dissolved in aqua regia, and tested with thallous nitrate (§ 26, *a*).

§ 146. Silver Alloys

a. Alloys of *silver* and *copper* readily dissolve in nitric acid. The solution can generally be immediately tested for *silver* with potassium bichromate. If very low in silver it must be tested with sodium chloride, or the silver must be precipitated on metallic copper. If the latter way be chosen, evaporate, redissolve in water, and add a small piece of copper. After an hour, immerse the copper in a large drop of caustic ammonia and brush off the precipitated silver.

b. Alloys of *silver* and *German silver* are dissolved in nitric acid, the nitrates are converted into neutral sulphates, and these are treated with metallic iron. The iron receives a coating of copper and silver, zinc and nickel remain dissolved. See Copper Alloys, German Silver (§ 140).

c. Alloys of *silver* and *aluminium* are likewise treated with nitric and sulphuric acids. If not too low in silver the solution of sulphates will, on cooling, yield crystals of silver sulphate (§ 87, *b*). Aluminium is afterwards precipitated with cæsium chloride. If very low in silver the alloy is scarcely attacked by nitric acid; in this case aluminium is dissolved in caustic potash, and the disintegrated residue in nitric acid.

§ 147. Platinum Alloys

Alloys of *platinum* are dissolved in aqua regia, and tested for platinum with ammonium chloride. Serious difficulty is encountered in dealing with alloys containing from 20 to 30 per cent of lead or silver. Such alloys are attacked sparingly by aqua regia, and not at all by nitric acid. They must be remelted with twice their weight of lead or cadmium to form an alloy fit for treatment with nitric acid.

For *Native Platinum* see the following section.

VI. EXAMINATION OF SOME COMBINATIONS OF RARE ELEMENTS

§ 148. Native Platinum (Platinum Ore)

The ore is heated with strong hydrochloric acid and

washed to remove *iron*. It is then treated with aqua regia at ordinary temperature.

a. If *gold* be present, the solvent becomes speedily light yellow, otherwise it will very slowly take a faint brownish tint. Concentrate till nearly dry, add a small drop of water and a grain of rubidium chloride, followed, if necessary, by a grain of thallous nitrate (§ 87, *c*).

b. For *osmiridium* its great hardness (= 7) would afford a valuable test but for small grains of spinel and zircon, frequently found in samples of platinum ore. A separation of the osmiridium from these minerals and also from platinum may be effected by fusing a sample of ore with four times its weight of copper and some borax in the blowpipe flame. The molten metal is run into a conical hole scooped out on the charcoal and kept in a liquid state for about two minutes. By this treatment platinum is alloyed with the copper, while the heavy osmiridium sinks to the point of the cone. The conic button of copper is dressed sideways and on the top till the file begins to grate on one of the hard grains. It is then dipped in strong nitric acid. The white grains of osmiridium will then come out strongly on a black background of powdery platinum. They are exceedingly hard, making deep scratches on glass and felspar.

c. If the alloy with copper or an analogous alloy with lead is treated with nitric acid, *palladium* is dissolved. It may be collected on a small piece of copper.

The powdery residue is treated with aqua regia. Platinum, iridium, and rhodium are dissolved with some copper or lead. If much iridium be present, part of it will remain as a crystalline powder. Ammonium chloride will precipitate reddish octahedra, composed of chloroplatinate and chloroiridate. Rhodium will remain in the solution with the base metal.

§ 149. Osmic Acid

a. If washed platinum ore is at once treated with aqua regia, the action is slow, even when assisted by heating. The solution may contain platinum, iridium, palladium, rhodium, gold, iron, and copper. Ruthenium remains undissolved with osmiridium. Osmic acid is volatilised. If it has to be traced, the treatment with aqua regia must take place in a small retort. The distillate is reduced to half its bulk by a second distillation, the vapours being absorbed by strong potash lye. Concentrate the alkaline liquid with some drops of alcohol and see if violet crystals separate out. If they do not appear, heat with strong nitric acid and a grain of potassium chlorate till two or three drops are driven off, which are again absorbed with caustic potash. A second concentration with alcohol will decide if osmium has been volatilised (§ 33, *b*). Some authors assert that a solution of potassium osmite will not bear heating; this statement does not hold good if an excess of caustic potash be present. In this case, concentration may be carried on with impunity at a boiling heat. The violet rhombic octahedra of osmite will vanish after some hours; they may be restored by addition of caustic potash and alcohol.

b. Fusing with a mixture of equal parts of caustic potash and potassium chlorate is an excellent method for extracting osmium from platinum ore and from osmiridium. If the sample be ground to fine powder in an agate mortar (a somewhat tedious operation), 5 centigr. of platinum ore are sufficient.

Begin with melting about 2 decigr. of caustic potash and 2 decigr. of potassium chlorate in a small porcelain crucible, add the metal, raise the heat to moderate effervescence and keep it up till the effervescence ceases. If osmium and ruthenium are present the molten mass will take a yellow or brownish colour. Dissolve in some drops of water and add strong nitric acid. If a strong smell of osmic acid is perceived, cover with a flat watch-glass, on which a drop of strong potash lye has been spread, and heat gently. The potash will be coloured yellow by osmate, which can be reduced to osmite by alcohol. *Iridium* is converted into black oxide, insoluble in all solvents.

c. By treatment with chlorine at a low red heat (Wöhler's method) all varieties of platinum ore can be decomposed, so as to yield soluble compounds. This method is expeditious and can be adapted for very small samples. Coarse-grained metal must be stamped, and ground to powder in an agate mortar. From two to ten centigr.

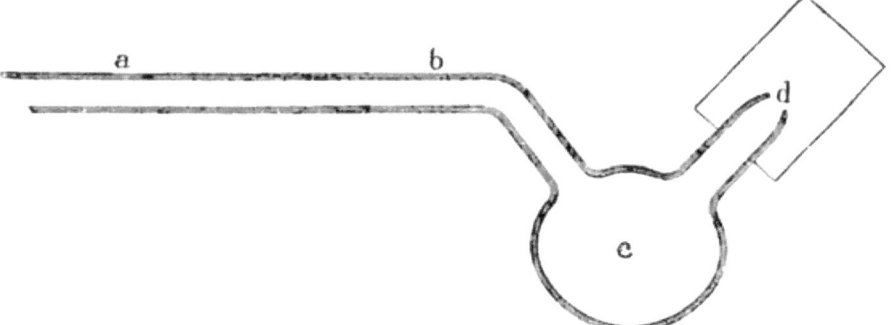

FIG. 84.—Chlorination tube. Full size. Width of the tube, 3 to 5 mm.

are intimately mixed with four times the quantity of calcined sodium chloride. The bulb of the chlorination tube is charged with potassium chlorate (from 0.5 to 2 decigr.), then the mixture of ore and sodium chloride is introduced at the opposite end and spread into a layer, filling the straight end of the tube to half its width. The straight part (a, b) is raised to a dull red heat, then a drop of strong hydrochloric acid is run in through d from a small pipette, a small cork is slipped on, and heating is continued from a to b till the colour of the mixture has become a saturated reddish-brown. If much iridium be present, the colour takes a darker shade. The operation will take from two to five minutes. If the greenish tint of chlorine is not seen between b and c, a drop of hydrochloric acid must be added. *Osmium* is betrayed by the smell of its volatile peroxide and by a red sublimate of osmium chloride at a. The tube is cut off at b, the contents are washed into a small crucible with some drops of water and left to settle. If a strong smell of osmic acid is perceived, heat with a drop of nitric acid and absorb the vapours of osmic acid with caustic potash (§ 149, b). Any considerable residue is dried, ground up with sodium chloride, and subjected to a second chlorination.

§ 150. Examination of a Solution containing Platinum, Iridium, Palladium, and Rhodium

This method is the result of numerous trials. It cannot lay claim to extreme delicacy, but it will be found expeditious and satisfies reasonable demands.

a. Evaporate the solution. If aqua regia has been used, dissolve in water; if chlorination has been employed, lixiviate with equal parts of water and alcohol and throw the cake of salt away when it is almost white. Drive the alcohol off at a gentle heat. Precipitate *platinum* and *iridium* with a slight excess of ammonium chloride. A reddish precipitate points to iridium. Evaporate, and wash with equal parts of water and alcohol.

b. The washing is concentrated with a small drop of aqua regia till the commencement of crystallisation is seen, when an excess of ammonia is added. If any considerable amount of *palladium* and *rhodium* be present, violet rods and rhombs (10 to 30 μ) of a chlorhodate of palladodiammonium will separate out. A slight heating will favour the crystallisation. If it turns out scanty, add a drop of hydrochloric acid, heat gently, and let the liquid dry up at ordinary temperature. Speedily great brownish bunches and crosses (150 to 300 μ) of the chloride of palladosammonium will appear, looking as if they were made of contorted shavings; around the border of the drop, rose-coloured rhombs and rectangles (15 to 60 μ) will be seen somewhat later. These are crystals of ammonium-rhodium chloride, sparingly soluble in a saturated solution of ammonium chloride.

FIG. 85.—Chlorhodate of palladodiammonium, ×130 : 1.

FIG. 86.—Palladosammonium chloride and ammonium-rhodium chloride, ×130 : 1.

c. If the test gives an unsatisfactory result, heat with an excess of hydrochloric acid, transfer the turbid solution to a piece of sheet zinc, heat slightly, rub the black precipitate of metallic palladium and rhodium off with a glass rod dipped in weak hydrochloric acid, and wash on a slide. Iridium that has escaped precipitation in *a* is also precipitated as metallic powder. The black particles are heated with nitric acid, which will dissolve *palladium* (§ 29, *a*). By heating with aqua regia *iridium* is slowly dissolved (§ 30, *a*). Any black dust remaining after repeated treatment with strong aqua regia is made into a paste with water and acid potassium sulphate. This is collected on a platinum wire and fused at a dull red heat. If rhodium be present, the mass will be red while hot; after cooling it will be yellow. It is dissolved in water, and tested with potassium nitrite and cæsium chloride (§ 31, *a*).

d. Platinum may be separated from *iridium* (precipitate *a*) by heating with water and oxalic acid. This treatment must be continued till the new crystals exhibit a pure yellow colour. The mother liquor, containing platinum tetrachloride and iridium trichloride, is boiled with aqua regia and evaporated. The residue is dissolved in a drop of water, and the solution is subjected to fractional precipitation with ammonium chloride. *Chloroplatinate* is precipitated first; later reddish crystals are formed, containing a considerable admixture of *chloroiridate*.

§ 151. Minerals containing Tantalic and Niobic Acids

Columbite, yttrotantalite, and allied minerals may be decomposed by fusing with acid potassium sulphate, or by fusing with caustic potash.

a. From a theoretical point of view another method, fusing with fluorides, looks most rational, leading by the shortest way to characteristic compounds of Si, Ti, Ta, and Zr. If this method be put into practice, however, serious difficulties are encountered, owing to variations in the composition of the fluosalts obtained in the dry way. If the fused mass be treated with sulphuric acid, silicon fluoride is driven off. The residue is almost identical with the product of fusion with acid potassium sulphate. Few minerals will resist this treatment.

b. Fusion with acid potassium sulphate is in current use in chemical laboratories. The operation must be conducted at a low red heat, which must be kept up till all is dissolved to a clear brownish liquid. If this should be found difficult, the cause will be volatilisation of sulphuric acid or want of solvent. Let the mass cool, add some acid sulphate and concentrated sulphuric acid, then heat again, at first cautiously, to avoid bubbling and spurting. The fused product is poured on a clean slab of iron or slate. Water will dissolve iron, manganese, aluminium, magnesium, and calcium, leaving tantalic and niobic acids in a state of comparative purity. This advantage is, however, outweighed by several inconveniences. Titanic acid is dissolved with the metals. This is sometimes in the operator's favour; but if small quantities of titanic acid are to be traced, it is not to the operator's advantage; and if zirconium be present it becomes very troublesome, part of the titanic acid remaining with the zirconium compound. Tungstic acid will remain with the tantalic and niobic acids; the same is true for zirconium, which will remain as an insoluble sulphate. If the cerite metals are present, their sulphates will form nearly insoluble compounds with potassium sulphate.

Fusion with acid potassium sulphate is of excellent service for separating *uranium* from the columbite acids in samarskite and analogous minerals. The solution is evaporated with an excess of ammonia; the residue is washed with water, and slightly heated with a strong solution of ammonium carbonate. The ammoniacal solution is tested with thallous nitrate (§ 59, *b*), or it is evaporated; the residue is dissolved in acetic acid and tested with sodium acetate (§ 59, *a*).

c. Calcination with sodium carbonate will yield soluble *sodium tungstate*, while titanic acid and the columbite acids are converted into insoluble sodium compounds. Sodium carbonate must be sparingly employed, just enough to form a caked mass, when the mixture is exposed to a bright red heat for half a minute. Any excess will give trouble in the final test. After short practice strong and pure solutions of tungstate are obtained, giving splendid crystallisation with thallous nitrate (§ 58, *c*), and a grain of caustic alkali. Precipitation as phosphotungstate is not to be recommended.

For a full microchemical examination the mineral must be calcined with caustic potash. If tungstic acid (and silica) have been extracted with sodium carbonate, two calcinations with caustic potash are necessary. The first portion of caustic potash serves to decompose sodium compounds. Subsequent treatment with water dissolves caustic soda and a little niobate. On the whole, two calcinations will give a better result than complete fusion with a great quantity of caustic potash, which makes the precipitation of the columbite acids tedious and incomplete. The first solution is generally coloured dark green by manganate. A little nitre may be added with the caustic potash, to bring all the manganese into this solution. It is precipitated by heating with a drop of alcohol, and the clear solution is filtered off, or the turbid liquid is evaporated and redissolved in water.

After the second calcination and extraction, the residue must be boiled twice with water, to extract all potassium compounds, especially the *tantalate*, which is far less soluble than the *niobate*. Both are precipitated as sodium compounds, which are almost insoluble in a strong solution of caustic soda. Sodium chloride has a slow effect, and is apt to fill the drop with cubic crystals of NaCl and KCl; sodium acetate will answer better. Begin by adding a strong dose of caustic soda; enforce its effect, if necessary, with sodium acetate. For more detail see § 46 and § 47. If a separation of tantalum from niobium is desired, evaporate, wash rapidly with cold water, add hydrochloric acid, wash, and transfer the columbite acids to a varnished slide for treatment with hydrofluoric acid (§ 47, *a*).

§ 152. Minerals containing Thorium, Titanium, Zirconium, Yttrium, etc.

Thorium, titanium, zirconium, the *gadolinite* and *cerite* metals may be present in the residue from the treatment with water mixed with a considerable quantity of iron oxide.

First Method—Treatment with Ammonia.—Thorium, zirconium, and yttrium may be extracted with a mixture of ammonia and ammonium carbonate, provided that Al and Ti are absent.

a. The residue from the extraction of the columbite acids is dissolved in strong hydrochloric acid; the solution is evaporated, and the dry chlorides are treated with the ammoniacal mixture at ordinary temperature. Part of the solution is tested for *thorium* with thallous nitrate (§ 38, *b*); the larger half is evaporated, the residue dissolved in weak nitric acid.

b. This solution is precipitated with acid potassium oxalate. The oxalates of thorium and yttrium come down immediately as a dense white powder; the double oxalate of potassium and *zirconium* (§ 37, *a*) will crystallise later along the border of the drop. It may be dissolved in hot water and recrystallised by concentration. The clear pyramidal crystals of $K_4Zr(C_2O_4)_4 + 4H_2O$ are to be sought for among the prismatic ones of acid oxalate of potassium.

c. To trace *yttrium*, the insoluble oxalates are gently heated with a solution of ammonium carbonate (§ 18, *b*). Small pyramidal crystals are produced, like those of calcium oxalate.

The residue from *a* is dissolved in hydrochloric acid. From the neutral solution sodium sulphate will precipitate binary sulphates of sodium and *cerite metals* (§ 15, *d*).

Second Method—Separation in Acid Solutions.—Both titanium and zirconium may be present.

d. Dissolve and evaporate as above; redissolve in water. Precipitate *cerite metals* with sodium sulphate (§ 15, *a*). The mother liquor is gently warmed with a grain of sodium sulphate, to ascertain if the precipitation of cerite metals has been complete. If a white powder is thrown down, ascertain under a high power if it exhibits the lenticular forms of sodium-cerium sulphate (§ 15, *a*), else test it for titanic acid with potassium ferrocyanide.

e. Boiling of the diluted mother liquor throws down *titanic acid*. It is left to settle. After a careful washing with dilute hydrochloric acid it is tested with a grain of potassium ferrocyanide. If pure, it is coloured an intense yellowish-brown; if not free from iron, it will turn a dirty green. Zirconium dioxide and zirconium sulphate are not stained. If the precipitate is too light for washing, it must be dissolved by heating with a small drop of strong sulphuric acid. Heating is continued till only a trace of sulphuric acid remains. After partial cooling a large drop of water is put on and at once heated to boiling.

f. The greater part of the titanic acid will be found adherent to the glass. To detect *zirconium* in a precipitate of titanic acid, testing with hydrofluoric acid or with ammonium fluoride (§ 37, *c*) is to be recommended. Enough material will generally remain dissolved, to be traced in the solution as *potassium-zirconium* oxalate. Thorium and yttrium are thrown down with oxalic acid; zirconium is kept dissolved by an excess of the reagent. It is precipitated from the mother liquor by addition of potassium oxalate.

g. The mixed oxalates of *thorium* and *yttrium* are calcined, the resulting oxides are treated with hydrochloric acid. Yttrium is extracted; it is again precipitated as oxalate and brought to the test with ammonium carbonate (§ 18, *b*). Thorium dioxide is finally dissolved by heating with strong sulphuric acid.

§ 153. Titanates and Zirconates—Vanadium

a. Zircon, cyrtolite, and *malacone* are decomposed by fusing with caustic soda at a bright red heat. The fused product is treated with water. Sodium silicate is dissolved, zirconate is left behind, freely soluble in hydrochloric acid. Minerals containing *zirconium* and *titanium* (polymignite, œrstedtite) are treated in the same way. *Tungstic* acid dissolves with the silica. *Uranium* (in euxenite) is detected in the liquid from which titanic acid has been precipitated by boiling (see Samarskite, § 151, *b*).

b. If titanium and zirconium are associated with *silicon* and *tin*, fuse with caustic soda, and treat with water, which will dissolve sodium silicate and almost all the stannate. Test a drop of the solution for tin (§ 35, *a*); to the remainder of the solution add a slight excess of hydrochloric acid, transfer to a varnished slide, and add ammonium fluoride. *Fluosilicate* and a little fluostannate of sodium will separate out, both of hexagonal shape. If any uncertainty remains, wash the precipitate and heat it on an ordinary slide with a solution of barium chloride. Both fluorine compounds are rapidly decomposed. The fluostannate yields shapeless grains, nearly opaque; the fluosilicate clear needles, like gypsum (§ 19, *b*).

Meanwhile the residue containing titanium and zirconium has been washed. It may be examined in two different ways. In hydrofluoric acid or in a solution of ammonium fluoride, acidulated with hydrochloric acid, it dissolves readily. The solution is tested with rubidium chloride. The *fluotitanate* (§ 36, *b*) will appear before the *fluozirconate* (§ 37, *b*). It is necessary to test also for iron and aluminium, to avoid being misled by fluoferrate and fluoaluminate of ammonium (§ 42, § 43, *b*). For this test a small drop of the solution in hydrofluoric acid is treated with ammonium acetate and an excess of ammonium fluoride.

The difficulty does not arise if the separation is effected without the use of fluorides. Dissolve in hydrochloric acid, add a little sulphuric acid, evaporate and boil with a large drop of water. *Titanium* dioxide is precipitated, adhering to the glass. Wash with hydrochloric acid, afterwards with a drop of water; test with a grain of potassium ferrocyanide. The strongest stain (brown or greenish, owing to a trace of iron) is generally found along the border of the evaporated drop, but even the thinnest films of titanium dioxide are sensibly stained. A little titanium dioxide, suspended in the solution, is of no consequence in testing for *zirconium* with acid oxalate of potassium (§ 37, *a*). The solution has to be concentrated before the reagent is added. This method is less elegant; but, on the other hand, it is more expeditious than the first.

c. Ilmenite and the other *titanic iron ores* are calcined with sodium carbonate and a little sodium nitrate at a bright red heat. Water will dissolve an excess of carbonate and at the same time *tungstate, stannate, chromate*, and *vanadate*. A separation may be effected by adding ammonium chloride. Amorphous *stannic oxide* is speedily precipitated; a great excess of the reagent will produce colourless lenticular crystals of *ammonium metavanadate* (§ 45, *a*). The mother liquor is acidulated with nitric acid and precipitated while hot with a small quantity of acetate of lead. The precipitate is thoroughly washed with hot water; it is tested for *chromate* of lead with a trace of caustic potash (see Ferrochrome, § 121, *b*) and with some more potash and thallous nitrate for *tungstic acid* (§ 58, *c*). The residue from the treatment with water is moistened with a small drop of hydrofluoric acid and heated with a drop of water. Under these circumstances very little iron is dissolved, while much titanium passes into solution as sodium *fluotitanate*. It is precipitated on an ordinary slide with nitre (§ 36, *a*). If very little titanium is present, more iron will be dissolved. In this case rubidium chloride must be employed instead of nitre.

§ 154. Tin and Tungsten

From the columbite acids *tin* may be separated in the same way as tungsten (§ 151, *c*). Water will dissolve tungstate, stannate, and a little niobate. Part of the tin remains undissolved. It might have been made soluble by employing caustic soda, but then a considerable quantity of niobate would have got into the solution. It is dissolved with the columbite acids after calcination with caustic potash, and remains in the solution from which sodium niobate and tantalate have been precipitated. By boiling this solution with ammonium chloride it may be precipitated with a little niobic acid.

With regard to testing for tungsten and tin two cases are to be kept apart.

a. Tungsten predominant. Fuse with caustic alkali, test the alkaline solution with thallous nitrate. To detect *tin*, evaporate with hydrochloric acid, extract stannic chloride with weak hydrochloric acid, concentrate till nearly dry, precipitate and wash with caustic ammonia. Dissolve the residue in hydrochloric acid and test with cæsium chloride (§ 35, *a*). As tungstic acid is apt to retain stannic oxide, it is dissolved in ammonia; any remaining film is tested as above.

b. Tin predominant. Test for *tin* in hydrochloric solution (§ 35, *a*). Metallic tin is twice evaporated with nitric acid; tin ore is calcined with sodium carbonate at a bright red heat, the calcined mass is extracted with water, and the solution is evaporated with nitric acid. This is done to convert the tin into an insoluble oxide. Wash to remove soluble salts (traces of chlorides would be troublesome), extract with a weak solution of caustic soda, concentrate, and test for *tungstic acid* with thallous nitrate (§ 58, *c*).

Traces of tungstic and molybdic acids may be detected in tin by dissolving in aqua regia, boiling with nitric acid to precipitate a great part of the tin, evaporating the solution and treating the residue as above.

§ 155. Compounds of the Cerite Metals, associated with Compounds of Iron, Aluminium, and Bivalent Metals.

a. Precipitation with oxalic acid (§ 15, *c*) answers well for solutions containing the cerite metals, together with aluminium, iron, or chromium. With bivalent metals it is apt to lead into errors, binary oxalates being formed, sometimes to such an extent that both metals are masked (cerite metals with magnesium or cadmium). Such fallacious oxalates can be analysed by calcination and evaporation of the oxides with formic acid. The residue is treated with cold water, which will leave formates of the cerite metals undissolved (solubility from 1 : 300 to 1 : 500).

b. With beryllium, magnesium, manganese, cadmium, and copper, precipitation of the cerite metals with sodium sulphate (§ 15, *a*) will give satisfactory results. With zinc, calcium, barium, strontium and lead it is otherwise. If zinc is present in any considerable proportion, sodium sulphate will not precipitate the well-known lenticular crystals (§ 15, *a*), but small needles, which might be taken for crystals of gypsum. Ammonia is of no use for extracting the zinc; for an effectual separation caustic soda must be employed (§ 131). Barium and lead may be separated from the cerite metals by precipitation with potassium bichromate (§ 19, *c*; § 22, *c*).

c. Calcium and strontium will be found more troublesome. Their sulphates form, with the sulphates of the cerite metals, granular compounds that cannot be distinguished under the microscope from barium sulphate. They must be converted into carbonates by fusion with sodium carbonate and careful washing of the fused mass, or into oxides by heating with a solution of ammonium oxalate, and subsequent washing and calcining of the insoluble oxalates (strontium sulphate is not decomposed).

The carbonates or oxides are treated with formic acid (§ 155, *a*). Calcium and strontium are precipitated from their solution in formic acid as tartrates (§ 21, *b*; § 20, *c*). For small quantities fractional precipitation is employed. The first grain of Seignette salt will precipitate the rest of the cerite metals as a dense white powder. A gentle heat helps to keep the tartrates of calcium and strontium dissolved. In the end well-defined crystals are obtained, which can be tested with sulphuric acid (§ 21, *a*).

§ 156. Chromium and Aluminium

From some silicates, by fusion with alkali and extraction with water, solutions are obtained containing very little chromate mixed with great quantities of silicate and aluminate. Such solutions are acidulated with nitric acid, heated, and precipitated with acetate of lead (§ 44, *b*). On the other hand, calcination of chrome-iron ore with alkali will sometimes give solutions containing a little aluminate. They are likewise acidulated with nitric acid and precipitated with acetate of lead while hot. From the filtered liquid lead is precipitated with a slight excess of sulphuric acid, the solution is evaporated, and the residue is washed with water. The clear washing is evaporated on as small a spot as possible. Any residue is tested in a minute drop of water with a grain of cæsium chloride (§ 42, *a*).

§ 157. Aluminium and Beryllium

a. Chrysoberyl and beryl are decomposed by heating with ammonium fluoride and sulphuric acid. To detect silicon in beryl another platinum spoon is put on for cover, moistened on the underside and cooled by keeping a drop of water in it. Phenakite (Be_2SiO_4) is treated in the same way. All ammonium sulphate and the excess of sulphuric acid must be driven off. The residue is dissolved in water and tested for aluminium with cæsium chloride (§ 42, *a*). When nearly all the aluminium has separated out as cæsium alum, potassium oxalate will show the presence of beryllium (§ 9). Neutrality of the solution, absence of ammonium salts, and a slight excess of potassium oxalate are essential conditions. The delicacy can be increased by adding a little acetate of magnesium or of zinc.

b. For traces of beryllium, a test, pointed out for sodium by Streng (see § 2, *b*), may be modified so as to suit the case of beryllium. The hydroxides of aluminium and beryllium are dissolved in caustic alkali, and the absence of zinc is established by testing with a trace of ammonium sulphide. Then enough acetate of ammonium is added to produce a slight precipitate. On boiling, beryllium and part of the aluminium are thrown down as hydroxides. The liquid is run off, the precipitate is dried and washed. It is then dissolved in acetic acid, the solution is concentrated as far as possible and tested with acetate of uranyl and a very small quantity of sodium acetate. The crystals of the triple acetate of uranyl, sodium, and beryllium are much larger than the yellow tetrahedra of the binary acetate of uranyl and sodium. They are besides almost colourless, even when reaching a diameter of 100 μ.

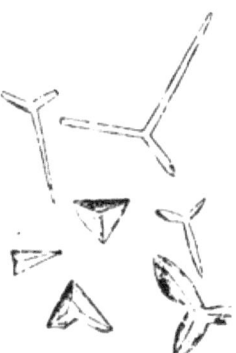

FIG. 87.—Triple acetate of beryllium, sodium, and uranyl, ×90 : 1.

Any excess of sodium acetate must be avoided. Finally the crystals are dissolved by heating with water and ammonium chloride. The solution is tested with ammonia and sodium phosphate, to establish the absence of magnesium.

§ 158. Beryllium and Magnesium

a. The separation of beryllium from magnesium is attended with some difficulty. It may be effected with caustic alkali, provided ammonium salts be absent. Evaporation ensures adhesion of the magnesium hydroxide to the glass, washing must, however, be performed with some care. The alkaline washing is evaporated with ammonium carbonate. Water will leave beryllium carbonate behind. If it is presumed that the quantity will be very small, addition of a little aluminium acetate to the alkaline solution will be found useful. It augments the bulk of the precipitate without any risk to the final test.

b. Another method is based upon the solubility of magnesium oxide in solutions of ammonium salts. A strong dose of ammonium chloride having been added, *beryllium hydroxide* is precipitated with caustic ammonia. A great part of the liquid may be immediately run off from the gelatinous precipitate. The latter is dissolved in hydrochloric acid, and purified by a second precipitation, followed by drying and washing. After this, it is dissolved in acetic acid and tested with potassium oxalate. The ammoniacal liquid is tested for *magnesium* with sodium phosphate. This test must also be applied to the drop containing the beryllium after it has been tested for this element, to ascertain if the separation has been satisfactory. The test for beryllium with potassium oxalate is, however, not injured by a slight admixture of magnesium. Taken by itself, potassium-magnesium oxalate is a very soluble compound, difficult to crystallise; by the presence of beryllium its behaviour is altered, a sensible quantity of potassium-magnesium oxalate being incorporated in the crystals of the analogous beryllium compound.

§ 159. Beryllium, Iron, and Manganese—Beryllium and Zinc

These combinations are found in *helvine* and *danalite*. Both minerals are easily decomposed by hydrochloric acid. The solution is evaporated, the residue is dissolved in a drop of water and treated with ammonia and hydrogen peroxide. After a slight heating, *beryllium* and *zinc* are extracted with ammonium carbonate. The solution is evaporated, the residue is dissolved in acetic acid and tested for *beryllium* with potassium oxalate. Potassium-zinc oxalate behaves in a similar manner to the analogous compound of magnesium (§ 158, *b*). Sometimes pale hexagonal plates will appear around the border of the drop, always a notable quantity goes into the crystals of potassium-beryllium oxalate. To discover *zinc*, oxalic acid is destroyed by heating with sulphuric acid, a drop of water is added, and a little thiocyanate of ammonium and mercury (§ 13, *c*). A trace of copper will be found useful for staining the thiocyanate of zinc and mercury.

§ 160. Beryllium, Aluminium, Iron, Yttrium, and Calcium

a. A combination, found in some varieties of *gadolinite*. Decompose the mineral with strong hydrochloric acid, evaporate, redissolve in water. *Silica* is left behind. Ammonia will precipitate beryllium, aluminium, iron, yttrium, and a trace of calcium. *Calcium* is precipitated as carbonate by heating the ammoniacal solution with ammonium carbonate. The film of calcium carbonate is slightly washed and tested with sulphuric acid (§ 21, *a*).

b. If oxalic acid or ammonium oxalate had been employed instead of ammonia, beryllium might have been overlooked, whereas subsequent treatment of the hydroxides with oxalic acid is very serviceable. *Yttrium* (like calcium) is converted into insoluble oxalate, beryllium, aluminium, and iron remain dissolved. The insoluble oxalate is washed and heated with sulphuric acid; the residue is treated with a solution of ammonium carbonate; a little calcium carbonate remains undissolved. *Yttrium* is detected in the ammoniacal solution by means of oxalic acid (§ 18, *b*).

c. The soluble oxalates are converted into sulphates by evaporating with sulphuric acid. Add a drop of water and some caustic alkali, evaporate and wash with cold water. *Ferric hydroxide* remains undissolved. The alkaline solution is evaporated with ammonium chloride; the residue is washed and tested for *aluminium* and *beryllium* (§ 157, *a*).

INDEX

Acids, testing for, 159
Ægirine, 190
Alkali (metals) separation, 161; separation from magnesium, 158
Alloys, 192; preparation of specimens, 192; colouring of specimens, 193; etching of specimens, 195; for bearings, 217; for *clichés*, 216; of copper, 205; of gold, 220; of platinum, 222; of silver, 221; for types, 215
Aluminium, 103; testing for, in rocks, 181, 187; in iron, 203; in bronze and brass, 209
Ammonia in water, 169
Ammonium, 121
Amphibole, 183, 190
Andalusite, 192
Anorthite, 188
Antimony, 115; in alloys of copper, 204; in alloys of lead and tin, 215
Apparatus, 12
Arfvedsonite, 190
Arsenic, 117; acid, 118; elimination of, 160; in alloys of copper, 204; in alloys of lead and tin, 215
Arsenious oxide, 117; in sublimated films, 150
Augite, 174, 183, 190

Barium, 63; group, separation of sulphates, 165

Basalt, 180, 189
Beryl, 238
Beryllium, 44; separation from magnesium, 240; from zinc, 241; from yttrium, 242
Biotite, 188
Bismuth, 113; testing for, in alloys of copper, 204; in alloys of lead, 217, 218; treatment of mixed sulphates of Bi, Ca, Na, 166
Boron, 102
Brass, 207; aluminium brass, 209; manganese brass, 212
Britannia metal, 214, 215
Bromine, 132
Bronze, 205; statuary bronze, 208; aluminium bronze, 209; silicon bronze, 210; Cowles's bronze, 211; manganese bronze, 212
Burners for microchemical operations, 19

Cadmium, 56; testing for, in alloys, 216
Cæsium, 35, 162
Calcium, 70; treatment of mixed sulphates of Ca, Ba, Sr, Pb, 165; of Ca, Bi, Na, 166
Camera-lucida, 14
Carbon, 101; combined with iron, 198
Carbonates in rocks, 176; in water, 168; precipitation of carbonates, 157

Cerite metals, 237
Cerium, 58
Chalcedony, 174
Chlorides, sublimation of, 151; precipitation of, 156
Chlorine, 131
Chlorite, 178
Chromite, 191
Chromium, 107; chromium and aluminium, 238; chromium in iron and steel, 201
Chrysoberyl, 238
Cobalt, 48
Colouring of metals, 193
Columbite, 228
Copper, 78; in iron, 203; alloys of copper, 205
Cordierite, 178, 182, 190
Cuprous oxide in copper, 204
Cyanite, 192
Cyanogen, 122
Cyrtolite, 233

DANALITE, 241
Didymium, 60
Disintegration of rocks, 181
Drawings of microscopic crystals, 14

ELÆOLITE, 179
Epidote, 175, 182
Erbium, 62
Etching of specimens of rocks, 175; of metals and alloys, 195
Euxenite, 233
Evaporation test, 146
Extraction of rocks with hydrochloric acid, 185; with hydrofluoric acid, 189

FELSPAR, 182, 183, 188
Ferroaluminium, 203
Ferrochrome, 201
Ferromanganese, 200
Ferrotungsten, 202
Filtration, 20
Fluorine, 135

GADOLINITE, 241
Garnet, 174, 182

German silver, 213
Gold, 83; alloys of gold, 220

HARDNESS, testing of minerals, 174; testing of metals, 193
Helvine, 241
Hematite, 174
Historical remarks, 1
Hornblende, 183, 190
Hydrochloric acid, examination of solutions in, 163; treatment of rocks with, 185
Hydrofluoric acid, treatment of rocks with, 181, 189

IDOCRASE, 182
Ilmenite, 191, 235
Introduction, xiii
Iodides, precipitation of, 156
Iodine, 133
Iridium, 89, 228
Iron, 105; ferric compounds, 106; ferrous compounds, 107; iron in brass, 208

JASPER, 174

LABRADORITE, 188
Lanthanum, 59
Lead, 74; in brass, 208; in bronze, 206; alloyed with tin and antimony, 214
Leucite, 178, 189
Lithium, 34

MAGNESIUM, 42; separation from alkali metals, 158; from beryllium, 240
Magnetite, 174
Malacone, 233
Manganese, 46; in brass, 212; in bronze, 212; in iron, 194, 200
Manganine, 213
Measuring of angles, 5; of microscopical objects, 14
Mercury, 81; mercuric compounds, 82; mercurous compounds, 81; testing for, by sublimation, 150
Metals, antifriction, 217; for *clichés*, 216; precious metals, 171, 219

INDEX

Mica, 178, 182
Microchemical analysis, aim of, 4
Microcline, **188**
Microscope, 12
Minerals, accessory, in rocks, 191; testing for hard minerals, 174
Molybdenum, 127

NEPHELINE, 179, **189**
Nickel, 51; in alloys of copper, 213; in iron and steel, 203
Nickeline, 213
Niobite, 228
Niobium, 110
Nitric acid, 121
Nitrites in water, 169
Nitrogen, 120

ŒRSTEDTITE, 233
Olivine, 174, 182, 189
Opal, 174
Ores, examination of, **170**
Orthoclase, **188**
Osmiridium, **223**
Osmium, 91; in platinum ore, 224
Oxalates, precipitation of, 158
Oxides, precipitation from solutions in nitric acid, **155**

PALLADIUM, 87, 227
Pewter, 214
Phenakite, 239
Phosphoric acid, elimination of, from solutions, 160; testing for, in rocks, 180
Phosphorus, 119; in bronze, 206; in iron, 199
Platinum, 85; platinic compounds, 86; platinous compounds, 85; alloys of, **222**; native platinum, 223; testing for, in ores and alloys, 172
Polymignite, 233
Potassium, 29; localised test for, 180
Pyrite, 174
Pyroxenes, 183, 190

QUARTZ, 182

REACTIONS, 29; table of, 137
Reagents, box for, 28; list of, 21
Rhodium, 89
Rhyolites, 179, 227
Rubidium, 37, 162
Ruthenium, 90
Rutile, 174, 191

SAMARSKITE, 229
Selenium, 124, 150, 156
Serpentine, 178
Silicon, 99; in bronze, 211; in iron, 199
Silver, 40; alloys of, 221; tracing of, in ores, 172
Slides, 17
Sodium, 31; mixed sulphates of sodium, calcium, and bismuth, 166
Specimens of metals and alloys, 192; of rocks, 174
Spinel, 191, 192
Staining of alumina, 181; of gelatinous silica, 176
Staurolite, 191
Steel, 195, 198
Strontium, 66
Sublimation tests, 149
Sulphates, examination of, 164; mixed sulphates of the barium group, 165; double sulphates of bismuth, 166; double sulphates of the cerite metals, 237
Sulphur, 123; in copper and its alloys, 204; in iron, 199

TALC, 178, 182
Tantalite, 228
Tantalum, 111
Tellurium, 125, 150, 155
Thallium, 38, 162
Thorium, 97
Tin, 92; stannic compounds, 93; stannous compounds, 92; tin in bronze, 206; tin, alloyed with lead and antimony, **214**; tin and arsenic, 216; tin and titanium, 233; tin and tungsten, 235; tin, zirconium, and silicon, 234
Titanium, 94, 233, 235

Tourmaline, 174, 191
Tungsten, 128; in iron, 202; tungsten and tin, 235

URANIUM, 130; allied with other rare elements, 229

VANADIUM, 109, 233
Volatile substances, testing for, 144

WATER, testing for, in solid substances, 153; microchemical analysis of water, 167

YTTRIUM, 62, 232, 241

ZINC, 52; in brass, 207; in bronze, 208
Zircon, 174, 191, 233
Zirconium, 96, 231

THE END

Printed by R. & R. CLARK, *Edinburgh*.

www.ingramcontent.com/pod-product-compliance
Lightning Source LLC
Chambersburg PA
CBHW031954230426
43672CB00010B/2147